True Gravity and
The Blueprint of The Universe

The Proof of Gravity's Cause

Dean E. Walker

This book is a work of non-fiction. Names of people and places have been changed to protect their privacy.

First published by AuthorHouse 10/28/04

ISBN: 1-4184-4199-6 (sc)

Library of Congress Control Number: 2004094293

Printed in the United States of America
Bloomington, Indiana

This book is printed on acid-free paper.

Photo Credits:
All photos and their narratives furnished by NASA through their media program. NASA has no other connection to this book, its contents, or its author, and has no opinion or judgments of it.

Front cover: Jupiter and four of its moons
Back cover: Saturn's system
Pictures of each body taken separately and combined.

Edited by:
Ginny Fernandez and Tracey Morin

authorHOUSE
Your Voice in Print

1663 Liberty Drive
Bloomington, Indiana 47403
(800) 839-8640
www.authorhouse.com

This book has been written
for the glory of God
and is dedicated to
all of my wonderful family.

Dean E. Walker

Table of Contents

PREFACE

THE PROOF OF GRAVITY'S CAUSE

Very little is known about the cause of gravity. Today's understanding of how the force of gravity is developed does not even answer the basic question of how it works.

Gravity as a so-called force of attraction explains nothing of how orbits are possible or how the heavenly bodies rotate. Nor does it explain why the planets are tipped on their axes or why they travel in elliptical paths. It can't explain why Venus rotates opposite to the other planets or why Uranus rotates at 97° to its path. It has no explanation for the different shaped galaxies of the Universe.

There are many different proofs in this book that gravity fields supply the gravity force and provide all of the answers to the above-mentioned phenomenon. There are countless examples of how the heavenly bodies act towards one another proving that gravity is not a force of attraction. A glaring proof that gravity is not a force of attraction is shown by the comparison of our moon's orbital force as it circles Earth to Io's orbital force as it circles Jupiter.

Io and our moon are almost twins in size and mass. Even the distances that they orbit from the centers of their planets are almost the same and would be if they were exact twins.

Io travels at 37,700 mph around Jupiter at a distance of 255,000 miles from Jupiter's center. Our moon travels at 2300 mph at a distance of 240,000 miles from Earth's center.

Because Io is traveling much faster in almost the same size orbit as our moon, it develops 253 times more centrifugal force than our moon does. Centrifugal force is the force that would cause the moons to break away from their orbits if the moons were not held inward by equal but opposite directed gravity force.

If gravity were a force of attraction not only would Jupiter's huge mass have to attract Io by 253 times as much as Earth would need to attract our moon; Io would have to attract Jupiter by 253 times more than our moon would need to attract Earth. But Io has about the same amount of mass as our moon has. The same amount of mass cannot have 253 times more attraction in one case than it has in the other. This is the boiler plate proof that gravity is not a force of attraction.

The sources' fields provide both the outward forces holding the satellites away from themselves and the inward force holding the satellites in orbit. A satellite takes its distance away from its source's center at the location in the field where both forces act equally on its size and mass.

All the stars are nuclear power plants that are fusing the protons of their atoms due to the huge pressure of the stars' gravity forces.

Our sun is a medium size star. It's continually stripping its matter apart and spewing its energies outward in the form of heat, light, and kinetic energy of the subatomic particles that are released from its atoms in the fusion process. It has been determined that there are over 150 different combinations of sub-atomic particles.

Subatomic particles are thousands of times smaller than their atoms and are therefore undetectable. It is proven that these tiniest parts of matter have mass and their mass is orbiting the stars, planets, and moons. The orbiting particles form waves. Each increment of distance from the source produces waves traveling at different speeds, depending on the distance from the center of the source. Those waves further from the source travel slower.

Joseph Taylor and Russell Hulse of Princeton University won a Nobel Prize in 1993 for proving that gravity waves are responsible for carrying energy away from a pair of binary stars as they react towards one another in the exact amount that the theory of relativity predicts.

This is the proof that gravity waves exist and that they do the work of transferring kinetic energy. Any force that transfers kinetic energy must be a direct force caused by the acceleration of mass. This proves that gravity waves contain mass. The force of gravity acts continuously, directing the kinetic energy of its waves of mass performing its task instantaneously. Gravity acts as a constant rate and force at any particular distance from the source.

It is shown in this book that Henry Cavendish's famous gravity experiment was also conducted in Earth's gravity field as well as being conducted in his laboratory. His findings that showed mass has an attraction for other mass has blocked all explanations of gravity's actual cause.

The force of gravity is not a force of attraction of one mass towards another but is due to the acceleration of mass in the gravity fields. The force of gravity is produced by the natural motions of the mass in the gravity fields by the expanding universe.

The Big Bang sent the universe into an everlasting expansion moving it into the frictionless void as it expands at millions of miles per hour. Initially matter of the Universe was unbalanced as it exploded outward making it whirl around. The Big Bang accelerated everything outward causing the spinning.

Now, the stars and planets are forced to rotate by two different momentums': the momentum given to them by the expanding universe and the momentum of their orbits. Any object acted on by two unbalanced forces, from two different directions, produces a third motion, the object's rotation.

Our planets obey the right-hand rule of engineering. The thumb of the right hand points upward in the direction of the expansion and the index finger points in the direction of their orbits, and the curled remaining fingers show the direction of rotation.

The rule applies to all the solar system's planets except for Venus and Uranus whose rotations have been interfered with and this is explained in detail in this book.

Planets and moons recycle the sun's field into their own gravity fields because the sun's field is weakened by the square of the distance away from it. The planets are millions and millions of miles away with Earth being 93,000,000 miles from the center of the sun.

Even the atoms of matter rotate due to these momentums given to their mass' inherent motions. If the Universe didn't expand, there could never be any gravity fields nor, could there ever be a force of gravity.

Subatomic particles are whirled around their stars, forming seas of undetectable mass. New material being released into the field makes room for itself as its centrifugal momentum forces the previous material into larger orbits as the whole field expands outward continuously.

Each of the stars is forcing millions of tons of the newly released particles outward every second.

The force of gravity is composed of two forces that are produced simultaneously by each wave. Waves of gravity take larger orbits, as they are forced centrifugally outward and accelerate their mass inward due to their curved paths. The shape and motions of the fields' waves of mass provides gravity's force.

Depending on which side of the wave that an object is located on, it is given the opposite effect by that wave as it is either forced inward by it or forced outward by it as the object is carried along at the speed of the field.

Anything to the inside of the curved field is accelerated inward in the way we are accelerated towards Earth as all of Earth's field is above us. Our moon is being held outward in orbit by the centrifugal advancing waves of Earth's gravity field while being balanced in its orbit by the inward curved waves of Earth's field beyond the moon.

Anything traveling in a curved path is being accelerated normally inward to that path as in the direction of spokes of a wheel towards its hub.

The timing of God's clock producing Kepler's clockworks is revealed in this book and is shown to be gravity's natural time constant. The time constant of gravity produces each star's gravitational acceleration as well as that of each planet and its moons.

The gravity time constant involves the amount of time each field orbits its source and produces each of their True Gravity Forces.

It's been only a hundred years from a time when no one could understand how we could ever fly in thin air. Today we all realize that the air is not that thin. Even though gravity fields are thousands of times thinner than air, they produce the mysteries of gravity.

Everything in the Universe is from the stars. Therefore, stars are also the sources of these undetectable fields of gravity.

CHAPTER 1

THE BEGINNINGS OF THE PROOF OF GRAVITY'S CAUSE

Gravity is the science of matter's force. Gravity is the force which controls the movement of all matter. It determines matter's location, speed, direction, size and shape.

No one has ever understood the cause of gravity.

We all know how it affects us. Most everyone is happy with knowing this and leaving the rest up to nature.

The Universe is in perpetual motion with the planets endlessly revolving around the sun, and the sun and stars revolving around their galaxies as the galaxies pass ever outward in an expanding Universe.

Just imagine the force required to cause the Earth to travel 66,700 mph in its orbit around the sun. Even more mind-boggling – imagine the force necessary to cause the sun to travel at 581,000 mph around the galaxy. The sun has 330,000 times more mass than does the Earth.

The force causing all of this is gravity and gravity is called the weakest of the natural forces. The sun's gravity force keeps Pluto in its orbit and Pluto is 3,660,000,000 miles away from the sun. As awesome as this is, it is dwarfed in comparison to our sun's orbit around our galaxy's center. The sun is 27,000 light years from the galaxy's center. Each light year is about 5,864,000,000,000 miles. It takes the sun 200 million years to make a trip around the galaxy.

I analyzed the force of gravity, determined to find its cause. I offer this book as proof of gravity's cause.

From the beginning, it was my intent to determine what causes the force of gravity. All the data for the planets and the sun has been tabulated. All the velocities, distances, specific gravities, sizes, masses, times of rotation, and length of years for each is known.

With everything known about the reactions of their matter, I felt it should be possible to determine the cause of the force producing it.

There have been many, many wondrous discoveries made about our Universe and our solar system. This history of these discoveries started even before Copernicus, who is known as the father of astronomy. It was he who first thought that the Earth traveled around the sun. Galileo helped to prove this concept as he studied the heavens with telescopes he had made. He could see that Jupiter's satellites were circling it.[1] This was compelling proof of how matter behaves.

Galileo also developed the formulas for the acceleration of gravity for matter towards the Earth. He used many different experiments and a lot of thought to develop his findings.

Johannas Kepler developed the three laws of planetary motion and joined with Galileo and Copernicus as the three who laid most of the foundation for Isaac Newton's great gravity Equation:

$$F = \frac{m_1 m_2 G}{d^2}$$

Isaac Newton developed his gravity equation in the mid 1600s and it is still in use today. Albert Einstein followed Newton in the early 1900s with his general relativity theory pertaining to gravity.

Gravity is truly the most remarkable phenomenon in all the Universe. Without it, there would be little else and certainly no witnesses.

Gravity is a lot like the chicken and the egg. Which came first as matter causes other mass to accelerate. How did the first mass become matter? How did it get its acceleration? These and many other answers await the reader.

KEPLER

Johannas Kepler's third law of motion shows that

$$\frac{t^2}{r^3} = K$$

The year (t) of each planet squared and divided by the cube of that planet's distance (r) to the sun equals the same constant for all the planets.[2]

Kepler was overjoyed to make this discovery. He said the solar system was a great clockworks and that he had discovered the mathematical proof of God's existence.

Using days for t and millions of miles for r, for each planet K works out to be .167. By carrying this work forward, it yields an amazing result.

Solving for t $\qquad t^2 = .167r^3$

By taking square root: $\qquad t = .41\,r\sqrt{r}$

Each planet's year can be computed by knowing only how far it is from the sun.

Earth*	$t = .41 \times 93 \times 9.63 =$	365 days
Mercury	$t = .41 \times 36 \times 6 =$	88 days
Mars	$t = .41 \times 141 \times 11.87 =$	686 days
Jupiter	$t = .41 \times 485 \times 22.02 =$	4,333 days

*The length of Earth's elliptical orbit needs to be considered to be a circle having a radius of 92,500,000 miles.

A year = $.41r\sqrt{r}$ is the rate the sun's acceleration causes any planet to make one complete orbit in days. This is true no matter what size it is or how far away it is. They all obey this simple formula.

Earths $r\sqrt{r}$ is $93 \times \underset{}{\overset{9.65}{\sqrt{93}}} = 889$. It says there are 889 days having 9.864 hours each of gravity's time or $\frac{9.864hrs}{24hrs} = .41$ and the .41 conversion to our 24 hrs/time of day is .41 x 889 = 364 days, approximately. Remember that $T = .41\ r\sqrt{r}$ has been derived by using time in days and r in millions of miles.

Earth's elliptical path causes problems for computations. Mercury and Venus orbits work even better in proving this point.

Mercury $\qquad 36\overset{6}{\sqrt{36}} = 216 \qquad 216 \times .41 = 88.5$ days/year

Venus $\qquad 67\overset{8.2}{\sqrt{67}} = 548.42$

548.42 x .41 = 225 days/year

CAVENDISH'S ATTRACTION

Henery Cavendish built an apparatus in 1798 that he had designed to show matter's force and its cause. It was composed of two large fixed iron balls and two moveable balls that were several times smaller.

The small balls were mounted on the ends of a steel rod that was suspended at the center of the rod by a thread from overhead. The small balls were balanced by the rod and thread. They were placed near the large balls, one on each end, to test if there was an attraction. If the small balls swing to the fixed balls by making the rod turn on the thread, it would show an attraction.

The rod turned and the small balls hit the larger fixed ones at each end. The rod was repositioned so that it would have to rotate in the opposite direction to make contact and it did.

Cavendish along with everyone else agreed that mass was attracted towards other mass. He developed the amount of the gravity constant needed to perform this task based on the mass of the balls.

<u>This experiment has hampered any and all understanding of the workings of gravity.</u> Everyone thought the experiment had been performed in Cavendish's laboratory. It was, however, conducted in the gravitational field of the Earth where the momentum of the expansion of the Universe and the galaxy's orbit worked to control the force of the gravity he was measuring.

The force Cavendish was measuring was not that of an attraction of mass to mass but was that of the acceleration of mass by the gravity field.

The gravity field swirled around the iron balls just as it does around the planets and their moons. The iron balls appeared to attract each other just as the moon appears to pull at the oceans.

Because the distance between the large balls and the small ones was so small it made no difference which side they were on. The field acted only to push them to the larger balls. Just as it makes no difference where we are on the Earth, the field is still all above us and only acts to accelerate us inward.

Yes, the size and mass of planets, moons, stars, and iron balls and everything else act in accordance to what its mass is. The field decides where it will all be.

Joseph Taylor and Russell Hulse of Princeton University won the Nobel Prize in 1993 for proving that gravity waves are responsible for carrying energy away from a pair of binary stars as they approached one another in accordance with the theory of relativity.

Gravity waves constitute gravity fields. The force of gravity is developed and delivered by gravity waves!

CHAPTER 2

IT'S A MECHANICAL FORCE

Einstein even showed that gravity could be replaced exactly by the acceleration of mass. He imagined an elevator in space. He said that a man inside the elevator in space was weightless when the elevator was not accelerating upwards. But, when the elevator was accelerated upward at $32.2'/sec^2$ (the same rate that gravitational acceleration is here on earth) the man would weigh the same there as his feet pushed against the floor of the elevator as he would on the Earth. Also, if the man dropped his hat, the floor of the elevator would rise to catch it at the same rate it would have dropped to the earth had he been standing here.

This is great. It shows how mechanical gravity really is. There is no attraction anywhere, only the force caused by the acceleration of mass.

On the Earth, the force of gravity acts in the other direction. The elevator is the gravity field of the earth and it accelerates us downward as all of the field is above us. We are to the inside of the curved field with none of its advancing field under us to hold us outward.

The further away from the Earth, the less field there is above us and the more advancing field behind us to hold us centrifugally outward. When the two become balanced for our mass and size, we would orbit at the speed of the field at that location.

Matter has mass. When mass is accelerated it becomes a force. A man in space is weightless, but he still has the same mass he has on Earth. The action of the field accelerates him inward equaling as much as it centrifugally pushes him outward. Here his mass is being accelerated towards the Earth and it gives him the force of his weight against the Earth. Mass in orbit is weightless because the outward acceleration of it equals the inward acceleration of it. The two balance each other canceling the acceleration portion of the force of gravity.

> $F = ma$ and $a = 0$ so, $F = o$, F being the force of the weight. Unaccelerated mass has no weight.

MOTOR OF THE MOON

Did you ever see a man ride a motorcycle around a circular wall at the fair? He seemed to fly around and around as he and his motorcycle were parallel to the ground.

There you could hear and see the motor and you see the curved walls. You understand the forces involved.

The moon is 240,000 miles away from us and is doing the same thing at 2,300 miles per hour.

We don't hear the motor and we can't see it nor do we see the wall, but they are both there.

The motor goes back to the beginning of time and roared louder than a billion billion nuclear blasts. It sent everything in existence into an everlasting whirling corkscrew motion straight out in every direction from the center of the blast. It was the BIG BANG!

The Big Bang is the motor of the moon and it powers the sea of orbiting subatomic mass of the Earth's gravity field, which centrifugally pushes the moon around and holds it outward.

The curved paths of the orbiting particles beyond the moon form the wall as their energy is being normally accelerated towards the Earth. This is the curved wall and its particles are so tiny that they can pass right through atoms with room for a thousand more beside them. So we can't see it. It's an invisible wall – a wall of dark matter if you prefer.

Very little is known about the cause of gravity. Today's understanding of how the force of gravity is developed does not answer even the basic question of how it works.

Gravity as a so-called force of attraction explains nothing of how orbits are possible or how the heavenly bodies rotate. Nor does it explain why the planets are tipped on their axes or how they travel in elliptical paths. It can't explain why Venus rotates opposite to the other planets or why Uranus rotates at 97º to its path. It has no explanation for the different shaped galaxies in the Universe or how black holes are possible.

There are many different proofs in this book that gravity fields supply the gravity force and provide all the answers to the above phenomena.

OUR MOON AND IO

The single greatest proof that gravity is not a force of attraction is by comparing our moon's force in its orbit to Io's, one of Jupiter's moons.

These two moons are almost twins in size, specific gravity and mass. Our moon travels at 2,300 mph around the Earth at a distance of 240,000 miles. Io travels at 37,700 mph around Jupiter at a distance of 255,000 miles.

We know that for a satellite to be kept in orbit its centrifugal outward force must be equally balanced by the inward acceleration of its mass. When the two forces are equal the orbit is established.

Normal acceleration for our moon's orbit is:

$$a_n = \frac{v^2}{r} = \frac{2300\,mph \times 2300\,mph}{240{,}000\,miles} = 22 m/hr^2$$

For Io:

$$a_n = \frac{37{,}700 mph \times 37{,}700 mph}{250{,}000 miles} = 5573.7\ m/hr^2$$

Io's orbital normal acceleration of its mass is $\frac{5573.7}{22}$ = 253 times more than our moon's and also its centrifugal outward force is 253 times greater than our moon's.

Jupiter's gravity field caused Io to produce 253 times more of the two forces than Earth's field causes our moon to produce around the Earth.

For this to take place, because of a force of attraction, it means not only would Jupiter's huge mass have to attract Io by 253 times as much as Earth attracts our moon, it also means that Io must attract Jupiter by 253 times more than our moon attracts Earth. Io must do this with about the same amount of mass our moon has. The same amount of mass could not produce 253 times more force of attraction in one case than it would in the other.

<u>The example of the two moons orbital forces is the complete proof that gravity is not a force of attraction!</u>

Io circles Jupiter in 1.77 days and our moon circles the Earth in 27.3 days at about the same distance away.

CHAPTER 3

A LOOK INSIDE OF THE GRAVITY FORCE

This book explains how gravity really works. It's a mechanical force produced exactly like all force is. It is the acceleration of mass. $F = ma$ is the equation discovered for force by none other than Isaac Newton and it applies to the gravity force just as it does to all other types of force.

The Universe is composed of matter all of which has either been part of a star, is now a star, or may become part of a star.

Gravity fields are composed of the parts, which used to be in the stars. They are now only parts of atoms or subatomic particles, many of them so tiny they are undetectable, even thousands of times smaller than the atoms they come from.

All stars are nuclear power plants, which are fusing the protons of their atoms due to the huge pressure at their centers.

Our sun, a medium size star, is continually stripping its matter apart from its atomic makeup and is spewing the resulting energies outward, powering our solar system. Including its massive field of gravity, the sun contains 99.9% of the mass in our solar system.

The vacuum of space is occupied by the teaming kinetic energy of undetected subatomic particles, which have been separated from their atoms by the nuclear reactions taking place in the stars.

Just as the stars, planets and moons obey the forces placed on them so does every subatomic mass in creation as they orbit their source.

The continuous expansion of the Universe provides the momentum powering gravity's force. Every particle of mass in the Universe, including yours and mine, are traveling outward at millions of miles per hour from the center of creation powered by the Big Bang.

If the Universe did not expand, there would be no gravity fields and no gravity force. The expansion is the motion of the galaxies moving outward. It is taking place because of a nuclear explosion, the Supernova, of what was the Universal Star or Egg.

The Universe itself has no gravity field, only its pieces do. It does not rotate nor does it orbit. Only its galaxies, stars, planets, and their satellites do.[3] Their curved momentums are combined with the linear expansion momentum and produce the mechanical force of gravity as it whirls the particles of subatomic mass around their bodies forming the fields of gravity.

Orbiting mass has a centrifugal outward force and it also has an equal inward force due to its curved path. Anything traveling in a curved path is being accelerated inward to that path. At any point on that path, it is being accelerated normally inward at 90° to the tangent of the curve at that point.

The inward force is from waves that are curved around the source by the preceding waves. Gravity waves advance, taking larger orbits as new waves push them outward from behind. This causes gravity to weaken as the waves are stretched and they also travel slower the further they orbit from the source.

The centrifugal advancing fields hold satellites outward in their orbits as they carry them around. The fields are curved fields of kinetic energy and when beyond a satellite, their orbiting mass forces the satellites to stay in their orbits as they crowd them around. The shape and motion of the entire field in back of them hold the satellites from escaping outward.

The combined force on a square inch of the Earth by the sun's orbiting gravity field is less than one millionth of a psi. However, the inward acceleration of the Earth's own field is responsible for producing the Earth's 32.2 ft/Sec^2 acceleration of gravity. It's this acceleration of the mass of Earth's field that has given Earth the shape of a sphere.

A steel ball drops in a vacuum tube at the same rate it drops outside the tube. It does this because the field of gravity is also inside the tube.

The vacuum tube acts the same as the so called vacuum of space. They are both filled with a sea of subatomic particles released by the destruction of the stars atoms by the nuclear reactions taking place in the stars. These are gravity's particles.

Subatomic particles are thousands of times smaller than the atoms of the material which the tube is composed of. The outside air is composed of atoms forming molecules and can't enter the vacuum tube through the walls. But, the gravity field passes right through the walls.

It has been said that if the nucleus of an atom were the size of the Sun there would be more empty space surrounding it than there is in our solar system. Thousands of subatomic particles could pass right through an atom at the same time.

A stream of water hits a rock in the middle of a brook and it curves around the rock surrounding it completely. Gravity fields act exactly like this with some of the particles hitting the nucleus of the atoms of matter. This produces a force so tiny on the tube that is immeasurable and therefore it is undetectable. However, this impacting of the particles of the suns field with an object the size of the Earth provides the force that carries it around in its orbit at the speed of the field. Orbits of satellites traveling into the oncoming gravity field can't be maintained as they lose speed and keep dropping into lower orbits. Orbits established by going with the flow of the gravity fields are sustained by the field.

Gravity is a constant force which works like a finely tuned clock. It controls the timing of the seasons, the nights and days along with the moons path and all are always on time.

The energy driving the system is supplied by the expanding Universe. It gives every particle of mass a corkscrew motion which whirls the gravity fields into orbit around the stars and planets, providing all the rotations and orbits in the Universe.

Gravity waves push mechanically outward and mechanically inward depending only on which side of the waves you are on. The same waves perform both tasks.

All of the mass ever created was created before the Big Bang took place. It changes into different combinations and even parts of atoms depending on what is happening to it, but by only using the standard parts which were all created before the Big Bang.

Matter of the Universe speeds outward from the center of creation on an endless journey. There is no friction to ever slow it down as it expands into the endless void.

Billions upon billions of galaxies comprise the Universe and each galaxy has a billion and some even have billions of stars.

Most of the stars were not created at the Big Bang as new stars are being born all of the time of the gasses in the gravity fields. The fields' swirling motion causes the gas to collect and it forms their shape.

The Big Bang sent everything into linear acceleration. It exploded straight outward in every direction from the center. All of the new pieces of the exploding sphere were unbalanced and were forced to spin fiercely as they were thrown accelerating outward. They became new stars, many of them having super gravity because they were spinning so fast.

The new stars passed through the billions of miles of the outer surroundings of the Universal Egg. Here they picked up the gasses and subatomic particles of the Egg's atmosphere and used them to help form their original gravity fields.

Matter was forced to rotate and to orbit due to its initial unbalance combined with the linear acceleration of its mass by the expansion.

The most important phenomenon taking place in the Universe is the nuclear reactions of the stars which produce the fields of gravity through their own destruction.

Because gravity is a direct force its field is actually pushing with its mass against everything else. Also, because the fields are composed of subatomic particles their mass must impart this direct force.

Waves of gravity are advancing, orbiting waves which are expanding outward into larger orbits. They are waves, like the ripples produced by a rock dropped into water, only gravity waves are also orbiting the source at the same time.

The only way that the tiny particles can impart their direct force is by making direct contact with other objects. But, because particles are so tiny that thousands of them could pass right through an atom at one time and not touch anything, it is imperative that the space in atoms already be full of them, forming the atoms own individual fields of gravity. This is the only way that subatomic particles can make a direct contact.

Therefore, atoms have their own fields of gravity of subatomic particles. These fields of gravity are what are released through the nuclear process of the atoms in the stars. The same tiny particles that once composed the fields of gravity of atoms join the huge fields of gravity of their stars and provide the answer to the theory of everything.

Just as Einstein said, the heavenly bodies take the path of least resistance, however, the particles of the gravity fields are also of mass and they obey the same principals that the larger satellites do.

Gravity fields are thousands of times thinner than any solid, liquid or even any gas. The Earth's atmosphere is filled with its own gravity field as it effortlessly orbits throughout the atmosphere. The closer the atmosphere is to the ground the heavier it is due to the acceleration of its own mass by the gravity field. At sea level the atmosphere weighs 16.4 pounds per square inch because of this.

The inward acceleration of the gravity field at the Earth's surface produces the maximum rate of gravitational acceleration of the Earth at 32.2 ft/sec^2. The further you go from the Earth, the less this acceleration becomes because less of the Earth's field is beyond you accelerating you inward.

The curved field of orbiting waves crowds the atmosphere inward forcing it into its shape and is accelerating its mass downward towards the center of the Earth. The weight of the Earth is accelerated by it, just as the weight of the sea is accelerated to its depth. The rock and the soil weight also accumulate creating huge pressures inside the Earth.

The inward acceleration of the orbiting field is produced by the orbiting mass of the field which accelerates the mass of the planets and their atmospheres inward.

All of the other three natural forces of matter could not exist without gravity. Gravity produces the rotation of the nucleus of all the atoms and causes their electrons and positrons to orbit them – just as gravity produces rotation and orbits for all other mass. Because an atom has mass, it is forced to rotate due to the expansion of the Universe and the orbit of its galaxy.

There could never be any type of natural force without gravity. The strong nuclear force and the weak nuclear force depend on the rotation of the atoms as does the electromagnetic force, so all three are nothing without gravity.

This is the theory of everything which Einstein sought for 15 years. "Because he chose to substitute the geometry that gravity created for its actual force"[4], he could not connect all of the natural forces in reality.

Gravity is, by itself, the answer to the theory of everything because without gravity there is <u>NOTHING.</u>

Gravity is the force which creates all the other natural forces.

The theory of reality is that the force of gravity causes actions all of which produce equal but opposite reactions.

There is a reason for every action and movement that we see taking place around us in the Universe. It's all due to the mechanical forces placed on mass by the effects of the expansion of the Universe.

Before the Big Bang, there was nothing. There were no stars and no Universe. The stars supply all of the energy in the Universe. Everything is of the stars. We are of the Earth whose minerals we possess and they were produced in a star.

Our star, the Sun, has a diameter of 868,000 miles compared to Earth's 7,910 miles. It supplies our heat, lights our day, and causes us to orbit it once a year in its field of gravity which the Earth recycles into its own field of gravity.

The sun, like all other stars, is a nuclear power plant. It sends all types of nuclear radiation outward including all of its deadly rays along with its sunlight. There are no fires at the stars. There is no oxygen to support the burning. Their own gravity force causes the stars to self-destruct as the huge force produced at their centers causes the protons of the atoms to fuse.

The fusion generates several different types of energy which is released to and through their fields. "The sun sends millions of tons of kinetic energy and mass outward every second."[5]

Kinetic energy is what the force of gravity is. Its mass times its velocity squared $KE = mv^2$. It comes about because of the momentum given to all mass as it is whirled outward and around by the expansion of the Universe. It causes the tiny particles of mass to be centrifugally orbited around their stars. <u>And the normal acceleration of their kinetic energy is the inward force of gravity.</u>

Stars have tremendous motions as they all are leaving the center of creation with their galaxies at millions of miles per hour. Many of them are orbiting their galaxies at over 500,000 mph like our sun does and possess rotational velocities at their surface from 1000 mph to 100,000 mph and more.

The expansion combines with the orbital and rotational motions to produce the stretching out of their paths, giving every mass a corkscrew whirling motion.

The fusion process generates force as it ejects kinetic energy into the orbiting fields. It pushes against the star to leave and it pushes against the field to enter.

The structure of the atom now has been determined to have over 150 subatomic changeable particle combinations, many having a tiny mass and several that have a + or – (plus or minus) charge.

Newly released particles make room for themselves in the field by pushing the previous waves of them outward into larger orbits. The energy of each wave is weakened as it is stretched out thinner as it is continuously being pushed into a larger orbit.

It's very hard to see a magnified atom and probably impossible to see a subatomic particle of an atom which is thousands of times smaller.

Each planet of the sun travels around it slower the further away from the sun it is. They are all traveling at the speed the sun's field is traveling in each of their orbits. The advancing field holds them outward and the curved field behind them holds them inward in a balance. The field, like anything that travels in a curved path, is being accelerated normally inward in the direction of spokes pointing toward the hub.

The mass of the orbiting waves is continually changing direction as it curves around in its orbit. This rate of change in direction of mass is acceleration. ($a_n = \frac{v^2}{r}$) It's this acceleration of mass which is its force of gravity. Just as Newton showed us force is $F = ma$.

It is the force in the force of gravity. Gravitational acceleration is produced when the mass of the gravity field is accelerated towards the source of the field.

The outward force of the waves is their mass being centrifugally thrown outward and around due to the motion and reaction of the expansion of the Universe. The preceding waves are curved causing all other waves to be curved due to the corkscrew motion of the expanding Universe.

Each wave pushes outward because it is thrown outward and each wave pushes around and inward because it is orbiting in a circular path, crowding anything inward which is inside of its path.

This is what gives gravity its two looks.

It holds us inward as all of the field is above us and it's all working to accelerate us inward. Yet it's busy keeping the moon held outward with the same waves as they advance towards the moon. This is how the moon just seams to hang there showing us it's possible.

GRAVITY IS A DIRECT FORCE

We can see by the lack of rotational speed of the closer planets and moons how hard they are trying to break away centrifugally. They stretch the source's field outward as they push on it behind them so hard that it slows their own rotating speed or it even stops their rotational altogether.

Stretching of the Earth's field is what our moon does. The moon is a huge satellite for the size of the Earth. As it tries to get away centrifugally it pulls hard on the Earth's field as it stretches it trying to escape centrifugally outward. This pulls the field up tight behind the Earth, causing pressure on the ocean behind the Earth, like stretching a rubber band around the two of them. It also releases the pressure on the ocean on the front side facing the moon.

The difference in pressure causes water to flow from the backside toward the front. It looks like the moon has an attraction for the mass but it doesn't. It's trying to get away.

A pressure of .43 psi rises one foot of water and a corresponding decrease on the side it flows to raising it another foot. An additional pound of pressure at sea level along with the corresponding pound of decrease in front creates a 4.6 foot tide (2.00 ÷ .43 = 4.6 ft).

The sun's field also causes tides on the Earth. The Earth similarly pushes on suns' field trying to escape centrifugally but the sun's field holds it, causing more pressure on Earth's backside pushing water to the front side.

CHAPTER 4

WHAT IS HAPPENING TO THE PLANETS?

In our Solar System all of the planets except two rotate towards the east. This is also the direction of all the orbits, counter-clockwise. Venus rotates backwards very slowly at only one revolution in 243 days, and Uranus has been tipped by over 90° to its path.

Uranus rotates 97° to its orbit toward the sun. It looks like a circular saw blade with its disc shaped field facing the sun as it would cut into the suns' field.

Uranus is very close to Neptune which is behind it. Neptune has more mass than Uranus and is much closer to Neptune than Uranus is to the sun. Neptune has flipped Uranus along with its field because of it. Uranus's field keeps it locked in this position with the gyroscope effect of the orbiting mass in its field. Uranus continues to feed its field from the sun's field because it is still powered by the expansion of the Universe.

When the planets first passed each other one behind the other in their orbits, their fields did not mesh and they tipped each other on their axes because of it.

When they are behind one another with their fields rotating in the same direction, they don't mesh. In order to mesh like gears, one needs to be rotating opposite to the other.

The further one's backside is moving to the right and the nearer one's front side is moving to the left as they both rotate in the same direction. When the fields try to pass through each other's fields, their masses are on a collision course and it kicks them so the disc-shaped fields tip away from each other and causes the planets to tip also.

The fields act like a spinning gyroscope by holding themselves and their planets in the new position. When the two planets line up one behind the other again, their fields are tipped away from each other and no longer interfere with one another. (See pg. 47)

It is interesting to note here that because the planets are tipped, their fields don't line up with the suns field either. This proves that the sun does all of the work of the orbits of the planets, showing once again how gravity works and that it is not a force of attraction, but is the force of the source's gravity field which controls their movements.

All planets are tipped to the plane of their orbits. They all tip upwards facing the sun except Venus which tips downward slightly at -2°. The Earth tips upward by +29°.

None of the other planets' fields make contact now because they are all tilted upwards and their fields are disc shaped just like our galaxy which has a disc-shaped field.

The sun's centrifugal force of its field pushes Venus hard against the sun's field behind Venus. Because Venus is traveling so fast at 78,000 mph on a curved path it stops Venus from rotating as the sun's field smothers it from behind. This is the same thing that happens to Mercury, our moon, and Io, Jupiter's moon.

Because Earth is behind Venus about 40% of the time Earth's field controls Venus's rotation. Venus is hampered by the sun and Earth's field is making Venus rotate backwards slowly.

One peculiarity of Io, Jupiter's moon, is that it never makes a complete rotation.[6] It wants to rotate slowly due to Jupiter but it never makes a complete rotation. It's because of its outside neighbor Europa, another one of Jupiter's moons. When Europa is behind it, it's made to start to rotate backwards. "Io travels twice as fast as Europa so Europa is behind it twice for each of Io's orbits. It takes Io only 1.77 days to make a trip around Jupiter and 3.55 days for Europa to do it."[7]

Each time Io starts to rotate correctly due to Jupiter's field, it is made to stop and rotate the other way by Europa's field. Europa is closer to Io than Jupiter is to Io. Also, Io is traveling so fast its centrifugal force against Jupiter's field mostly smothers its rotation. Europa is beyond Io and the back side of Europa's field moves to the right as the front side of Jupiter's field is moving to the left as they both rotate counter-clockwise.

Io only twists back and forth, not completing any revolution.[8] This is about the same effect the Earth has on Venus as it causes Venus to rotate backwards to all the other planets.

Another thing about Io is that it is losing some of its sulfur to Jupiter's field.[9] The sulfur leaves Io at different rates of speed depending on the strength of its own gravity field. At times it loses sulfur up to 20,000 mph to Jupiter's field. At other times none of it seems to even be leaving, indicating its varied field strengths. This proves that Io has a field of gravity because it loses strength and then gains it back, depending on what is happening to its field.

Europa's field interferes with Io as it has the opposite effect on it as Jupiter does. The sulfur forms a ring around Jupiter similar to the rings around Saturn. Each particle of dust or sulfur acts like a tiny moon in an orbit forming the ring. The ring shows the strength of Jupiter's field as it orbits around its equator.

CHAPTER 5

EVERYDAY OCCURENCES

The force of gravity requires a field of gravity. Just ask the man in the moon. His existence depends on it just as ours does.

Einstein once saw a man fall from a ladder and he ran to ask him if he had felt any force pushing him down. Falling is being unsupported. If he had fallen in space he might have fallen away instead of inward because the field might have been stronger in centrifugal force than in normal acceleration at that location and would not support him in that direction.

When a sink is drained the falling water is whirled around and outward against the vertical drainage pipe by gravity. When a toilet is flushed the swirling of the water is caused by the same force.

When cold air above warmer air falls down through it, it swirls outward as it drops producing a vacuum in its center and the winds rage around the funnel shape produced by the tornado. It forms a funnel because the air pressure around it increases the nearer it gets to the ground causing it to neck down.

All of these examples rotate counter-clockwise in the Northern Hemisphere of the world and clockwise in the Southern Hemisphere. They all rotate to the east, the same as the Earth is rotating and orbiting. But, because the Southern Hemisphere is upside down, it produces a clockwise motion.

Anything dropping through the Earth's orbiting gravity field is made to swirl because the linear acceleration of its mass is also being accelerated sideways by the orbiting field of gravity.

All of the previous examples of falling swirling water and air are everyday occurrences. A gravity force which would only attract matter towards other matter would be unable to produce any of them any more than it would be able to cause the moon to orbit the Earth.

Anything falling to the Earth is forced to turn just as the stars and planets do as they move outward with the expansion and are carried around by the orbiting fields of their galaxies.

As reported on TV news: A small amount of water was spilled in a spaceship while on the way to the moon. The water quickly formed tiny spheres and floated in mid-air. The spheres appeared to be perfectly round and were much larger than just drops of water. They were said to be facinating to watch, all shiny and sparkling while they were spinning around and around their own axes' like tiny minature planets as they orbited with the spaceship.

When our flag was planted on the moon it rippled. It didn't ripple in the wind as there is no atmospher at the moon. Could it be that it was rippled by the moons orbiting gravity field?

HOW IT MIGHT HAVE HAPPENED

All of the heavenly bodies were surrounded by subatomic mass of the Universal Egg when it exploded into the Super Nova of the expansion of the Universe.

In the beginning there was no Universe nor were there any subatomic particles or atoms, molecules, gas, dust, or stars. There was only an endless void of darkness in absolute cold.

There are several different opinions as to how the Universe came to be. However, I choose to write about my own.

I believe that God created his Universe. He built it with the tiniest of particles of matter. These particles became atoms and atoms became molecules and the molecules became gas.

The gas held together in a cloud by its molecular attraction and began to bear against itself from every direction for support, forming a sphere. The sphere grew and grew for a billion years causing a huge pressure at its center as all the weight acted from every direction towards the center.

The sphere did not rotate as we know the Universe does not rotate.

Protons began to fuse together because of the pressure. They released energy and subatomic particles to its outer surroundings.

It has become a star, the star of the Universal Egg. Because it didn't rotate it was unaccelerated. This is why it was able to get so huge as to include the whole Universe.

The subatomic particles released were different than the ones forming the gas and they did not combine with each other.

After the Egg and its pressure had grown for a total of about 5 billion years, it exploded.

The explosion was the largest nuclear explosion of a Super Nova ever recorded as it is still being recorded 10 billion years after the blast.

The center of the Egg was molten and formed billions of huge molten globs of mass when it exploded. They were sent whirling out of balance in every straight diverging outward direction.

It resembled a fire works rocket that explodes and grows into a larger sphere of shining stars. Only the Universe keeps expanding into the frictionless void and fills it with what there is.

The globs of molten mass were out of balance causing their wobbly rotation at first. They were accelerated outward so fast that they were made to spin very fast becoming stars. Their rotation was faster than normal stars today. Their shapes soon became spheres as their surfaces smoothed.

These new stars were passing through billions of miles of the outer layers of the Egg. They picked up gas and subatomic particles as the corkscrew motion of the expansion whirled the particles around them, forming their initial gravity fields.

The gas is still forming new stars in the galaxies' fields due to the swirling motion of the expansion which gathers it in and shapes it.

The egg had hatched and its litter was growing and expanding with tremendous acceleration. It had exploded and was leaving the center. Nothing is left there as everything pushed against itself and left.

The galaxies have their own fields, the stars all have their own fields, and the planets and moons recycle their star's fields in their own location forming their own fields which are being continuously re-supplied by their star and galaxy.

Universal expansion continues at millions of miles per hour. It will never stop expanding because the Universe has no gravity field of its own. Everything in it is leaving the center of creation at millions of miles per hour. There is no place to get a toehold to ever slow it because gravity is not an attraction of one mass toward another. If it was, the expansion would slow, not accelerate.

As we observe the galaxy near us going out beside us we have about the same speed out but we are still separating sideways at a much slower rate. It's like we are each going out opposite sides of a wedge of pie from the center outward.

Our Earth and sun with our galaxy are all leaving the center of creation with the same tremendous velocity.

The galaxies hold themselves together with their fields but they spread apart from other galaxy neighbors as they go outward radially from the center of creation.

We orbit the galaxy at 580,000 miles per hour with the sun taking us with it. One trip around our galaxy takes us 200 million years. We orbit the sun at 67,000 mph and our round trip takes a year. Knowing that we orbit the galaxy at 580,000 mph and feel nothing; it becomes easier to comprehend the tremendous rate of the expansion and the fact that we feel nothing of it either.

Scientists believe that the stars furthest from us are leaving us at a faster rate than the nearer ones. The stars furthest away from us would be located directly on the opposite side of the Center of Creation and would be traveling just as fast as we are, only in the opposite direction. That's why they appear to be leaving the center faster as one half of the speed is our own in the other direction.

There are many scientists who believe that the Universe is flat, and it might be. If the Universal Egg was always falling, the speed of the explosion of its upward motion would be canceled by the downward speed of its falling. This could have produced a flat Universe, even though it is produced from a round sphere.

CHAPTER 6

SUPER-GRAVITY

The huge rotational speed of the White Dwarf star is thousands of revolutions per second.

How can it hold together and not fly all apart due to the huge centrifugal force wanting to make it do just that? Everyone says it does it because of its huge gravity force. Yes, but what kind of force is this that works backwards?

In order for it to have a huge weight or force, it must be accelerated inward. But centrifugal force acts outward and spinning objects fly apart due to that, especially heavy ones.

This is the next tremendous proof of the existence of gravity fields.

For the weight of the star to be gigantic it must be pushed inward against itself from all directions. And for the field to be strong, it must be centrifugally pushed outward just as much. The White Dwarf star has a tremendously strong gravity field.

The centrifugal force that would cause it to fly apart is used to push its strong field outward. In doing this, the strong field acts inward equally. The inward force of the fields normally accelerated kinetic energy gives the White Dwarf its tremendous weight as the entire field's inward force acts to crush it. This is the force which produces the spherical shapes.

The same thing applies to black holes where a teaspoon of matter weighs many tons. It weighs so much because of its acceleration. It's the acceleration of its mass that is the force of its weight.

A black hole is the proof that super gravity takes all of its matter's kinetic energy and turns it into the gravity force.

Super gravity is caused by super rotational velocity. It is produced in stars at the end of their lives when they collapse inward onto themselves.

Before a normal star collapses, it loses internal pressure and expands into a Red Giant star 100 times its normal size. A star of a million-mile diameter becomes a Red Giant with a 100 million mile diameter. It is no longer white-hot and it turns red as it cools down.[10]

When the star was shining brightly it has a surface velocity of rotation of around 4000 miles/hour like our sun.

Red Giant stars don't live long. As their internal pressure fails, the star falls inward onto itself, forming a 20,000 mile sphere of the White Dwarf star.[11]

When the original star was a bright star it had a million-mile diameter. Now it is compressed into a star of only a 20,000 mile diameter.[12]

The White Dwarf star formed now shines 100's of times brighter than ever before because it is now rotating at 1000's of revolutions per second.[13] The particles still retain the rotational velocity of the Red Giant. Nothing has happened to slow the rotational velocity of the particles of the Red Giant or its field's particles.

The Red Giant star made only one revolution in days of its one hundred million mile diameter size. But now its particles are all contained in a 20,000 mile-diameter White Dwarf star, and the rotational velocity overwhelms the tiny White Dwarf, making it produce a tremendously strong gravity field.

The field also has collapsed from the Red Giant's 100,000,000 mile diameter inward to a 20,000 mile diameter. Its velocity is retained and the field orbits at 1000's of revolutions per second also. Remember, that the field of the Red Giant circled it at 9.864 hours and now still has the same velocity.

This is the super rotational velocity that causes super gravity and produces a super gravity field that keeps it all together. Black holes come into being in the same manner.

Older stars contain much less mass than they did when they were younger. A large part of their mass has been converted into different types of energy.

The amount of mass a star has determines the size and strength of its gravity field. The field's strength becomes weaker because the amount of the stars mass has shrunk. When the field becomes weak enough it will no longer force the gases of the star inward, and it swells outward, going out with its advancing field.

A force of gravity developed by a force of attraction could never allow a Red Giant to evolve. Its force would only act towards itself and the star could only be smaller, never larger, as it looses internal pressure.

The existence of Red Giant stars is in itself proof that gravity is not a force of attraction.

BLACK HOLES

Black holes are a "for real" phenomena. The centers of some galaxies have black holes containing a billion star masses and they can't even be seen.[14] But, we know they are there because their force is necessary to control their 200 light year diameter galaxies.

It is said that the escape velocity of a black hole is so large that light can't leave it. But its gravity field has to leave it because it's responsible for the orbits of the stars in its galaxy.

Black holes keep shrinking because of their field's strength. A shrinking black hole works like a spinning ice skater who pulls in his arms for a spin. The closer mass is to its center, the faster it rotates.

When the black hole makes more revolutions per second, it increases its strength, which further shrinks it causing it to pick up more speed and on and on.

A single star that produces a black hole with nothing around it to feed its appetite is the purest example of super gravity. It keeps shrinking its mass and growing in gravity's energy as its mass produces the kinetic energy.

It is said that gravity is the weakest force of the four natural forces of matter. But, in its super state gravity overcomes everything about matter as it bullies its mass, turning it into its own energy.

MANNED SPACECRAFT CENTER, HOUSTON, TEXAS

APOLLO 17 VIEW OF EARTH -- A fantastic view of the sphere of the earth as photographed from the Apollo 17 spacecraft during the final lunar landing mission in NASA's Apollo program. This outstanding photograph extends from the Mediterranean Sea area to the Antarctica south polar ice cap. Note the heavy cloud cover in the Southern Hemisphere. Almost the entire coastline of the continent of Africa is clearly delineated. The Arabian Peninsula can be seen at the northeastern edge of Africa. The large island off the southeastern coast of Africa is the Malagasy Republic. The Asian mainland is on the Horizon toward the northeast.

FOR RELEASE: December, 1972
PHOTO NO. 72-HC-928
72-H-1578

Jupiter, its Great Red Spot and three of its four largest satellites are visible in this photo taken Feb. 5, 1979, by Voyager 1. The spacecraft was 28.4 million kilometers (17.5 million miles) from the planet at the time. The innermost large satellite, Io, can be seen against Jupiter's disk. Io is distinguished by its bright, brown-yellow surface. To the right of Jupiter is the satellite Europa, also very bright Callisto (still nearly twice as bright as Earth's Moon), is barely visible at the bottom left of the picture. Callisto shows a bright patch in its northern hemisphere. All three orbit Jupiter in the equatorial plane, and, appear in their present position because Voyager is above the plane. All three satellites show the same face to Jupiter always – Just as Earth's Moon always shows us the same face. In this photo we see the sides of the satellites that always face away from the planet. Jupiter's colorfully banded atmosphere displays complex patterns highlighted by the Great Red Spot, a large, circulating atmospheric disturbance. This black-and-white photo was taken through a violet filter. Jet Propulsion Laboratory manages and controls the Voyager project for NASA's Office of Space Science.

FOR RELEASE: February 21, 1979
PHOTO NO. 79-HC-65
79-H-80
P-21083
1-22 B/W

This photograph of the Sun, taken last December 19, by NASA's Syklab4, shows one of the most spectacular solar flares (upper right) ever recorded, spanning more than 588,000 kilometers (367,000 miles) across the solar surface. The last previous picture, taken some 17 hours earlier, showed this feature as a large quiescent prominence on the eastern side of the Sun. The flare gives the distinct impression of a twisted sheet of gas in the process of unwinding itself. Skylab photographs such as these may provide clues to the mechanism by which such quiescent features erupt from the Sun. In this photograph, the solar poles are distinguished by a relative absence of super-granulation network, and a much darker tone than the central portions of the disk. Several active regions are seen on the eastern side of the disk. The photograph was taken in the light of ionized helium by the extreme ultraviolet spectrohelograph instrument of the US Navel Research Laboratory.

FOR RELEASE	June 7, 1974
PHOTO NO.	74-HC-260
	74-H-434

This full disk picture of Venus was taken by the Pioneer-Venus Orbiter on Feb. 10, 1979. The Orbiter so far has sent back some 347 images and been circling the planet for 5 months. This image was taken at 65,000 km (40,000 miles) from Venus. It perhaps shows an arm, in the southern hemisphere, of the Y with little indication of a northern hemisphere counterpart. This image shows very irregular versions of the dark Y.

The last image was taken on May 2, 1979 and the next will be taken on June 2, 1979. The month of May will be devoted solely to the gathering of polarimetry data This phase angle or the variation of illuminations of Venus, has never been seen from Earth.

The Orbiter began its orbit of Venus on December 4, 1978. The Pioneer Venus Multiprobe, which split into four probes and a bus, entered Venus' atmosphere on December 9, sending back extensive data before the planet's searing atmosphere destroyed the probes on the surface. The Multiprobe and Orbiter spacecrafts are managed by NASA's Ames Research Center, Mountain View, CA. The spacecraft were built by Hughes Aircraft.

FOR RELEASE: May 28, 1979
PHOTO NO. 79-HC-223
79-H-298

THE SPIRAL GALAXY M100

An image of the grand design of spiral galaxy M100 obtained with NASA's Hubble Space Telescope resolves individual stars within the majestic spiral arms. (These stars typically appeared blurred together when viewed with the ground-based telescopes.)

Hubble has the ability to resolve individual stars in other galaxies and measure accurately the light from very faint stars. This makes space telescopes invaluable for identifying a rare class of pulsating stars, called Cepheid Variable stars embedded with M100's spiral arms.

Cepheid's are reliable cosmic distance mileposts. The interval it takes for the Cepheid to complete one pulsation is a direct indication of the star's intrinsic brightness. This value can be used to make a precise measurement of the galaxy's distance, which turns out to be 51 million light-years.

M100 (100th object in the Messier catalog of non-stellar objects) is a majestic face-on spiral galaxy. It is a rotating system of gas and stars, similar to our own galaxy, the Milky Way. Hubble routinely can view M100 with a level of clarity and sensitivity previously possible only for the very few nearby galaxies that compose our "Local Group."

M100 is a member of the huge Virgo cluster of an estimated 2,500 galaxies. The galaxy can be seen by amateur astronomers as a faint, pinwheel-shaped object in the spring constellation Coma Berenices.

This color picture of Mars was made from three frames shuttered nine seconds apart by the Viking 1 Orbiter on June 18. Each of the three pictures was taken through a different filter—red, green and violet. Color reconstruction was done at the Image Processing Facility of the U.S. Geological Survey in Flagstaff, Ariz. Just below center of picture and near morning terminator is the large impact basin Argyre. Interior of the basin is bright, suggesting ground frost or a ground haze. Bright area south of Argyre probably is an area of discontinuous frost cover near the south pole. The pole, itself, is in the dark at lower left. North of Argyre, the "Grand Canyon" of Mars, called Vallis Marineris, can be seen near the terminator. Marking elsewhere on the planet are mostly due to differences in brightness; however, color differences are present, suggesting compositional differences. Area at top is the eastern side of the Tharsis volcanic region and is bright because of cloud activity.

FOR RELEASE: Filed: June 25, 1976
PHOTO NO. 76-HC-624
76-H-482
Viking 1 – 18
P – 16870 (color)

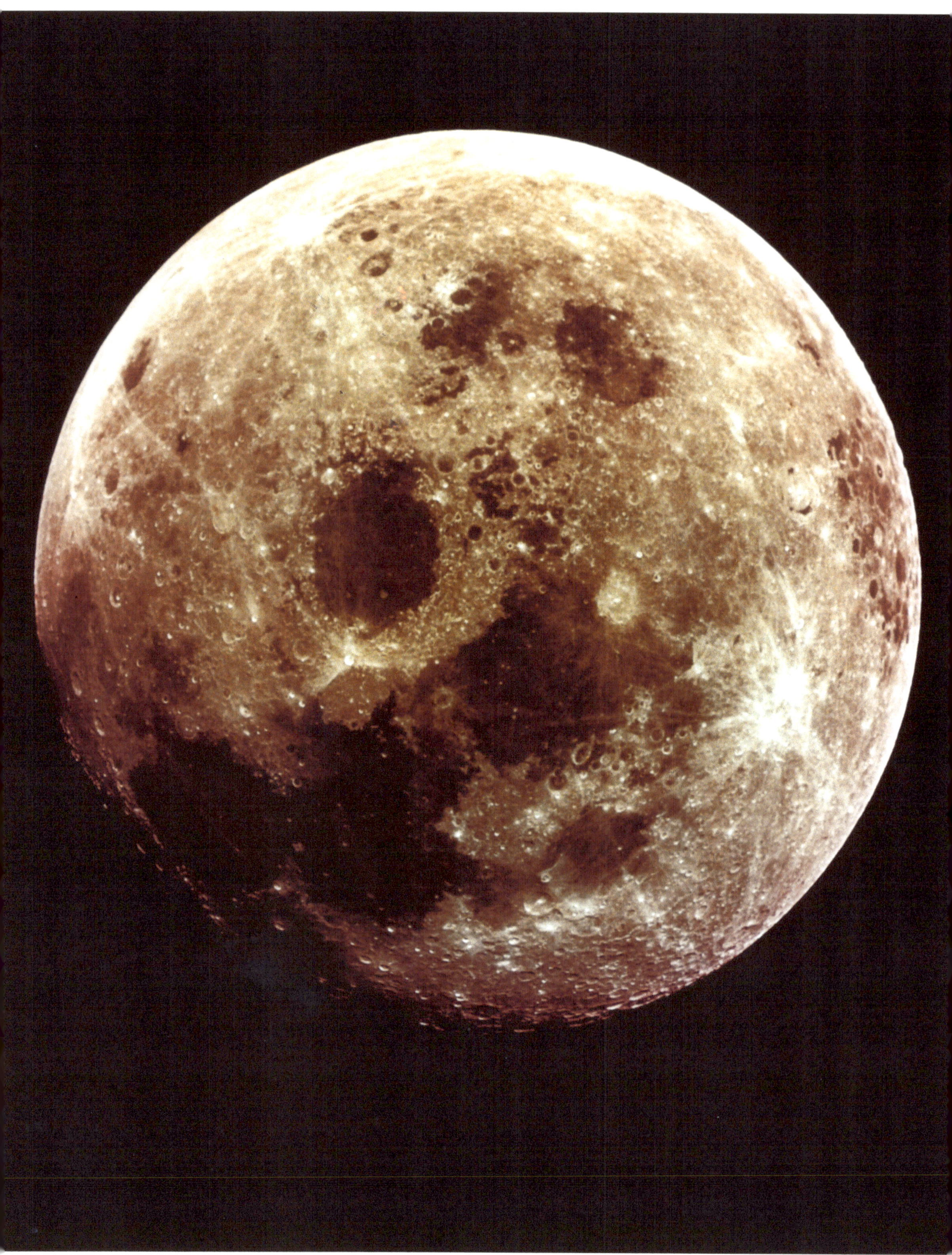

MANNED SPACECRAFT CENTER, HOUSTON, TEXAS

APOLLO 11 VIEW OF THE MOON – This outstanding view of a full moon was photographed from the Apollo 11 spacecraft during its transearth journey homeward. When this picture was taken the spacecraft was already 10,000 nautical miles away. Aboard Apollo 11 were Astronauts Neil A. Armstrong, Michael Collins, and Edwin E. Aldrin. Jr.

FOR RELEASE: July 21, 1969
PHOTO NO. 69-HC-901
69-H-1374
AS11 – 44 - 6667

This true color picture was assembled from Voyager 2 Saturn images obtained August 4 from a distance of 21 million kilometers (13 million miles) on the spacecraft's approach trajectory. Three of Saturn's icy moons are evident at left. They are, in order of distance from the planet: Tethys, 1,060 km. (652 mi.) in diameter; Dione, 1,120 km (696mi.); and Thea, 1,1530 km (951 mi.). The shadow of Tethys appears on Saturn's southern hemisphere. A fourth satellite, Mimas appears on the planet about three quarters of an inch directly above that of Tethys. The pastel and yellow lines on the planet reveal many contrasting bright and darker bands on both hemispheres of Saturn's weather system. The Voyager project is managed by NASA by the Jet Propulsion Laboratory, Pasadena, California.

FOR RELEASE: August 24, 1981
PHOTO NO. 81-HC-520
81-H-582

Eighteen pictures, taken at 42-second intervals by Mariner 10's two TV cameras, were computer-enhanced at the Jet Propulsion Laboratory and fashioned into this photomasaic of Mercury. The pictures were taken during a 13-minute period when Mariner was 200,000 kilometers (124,000 miles) and six hours away from Mercury on its approach to the planet March 29. About two-thirds of the portion of Mercury seen in this mosaic is in the southern hemisphere. The cratered surface is somewhat similar to the cratered highlands on the Moon. Largest of the craters are about 200 kilometers (124 miles) in diameter. Illumination is from the right.

FOR RELEASE:	March 31, 1974
PHOTO NO.	74-H-239
	Mariner 10 – 27 (Mercury)
	P - 14470

Solar System – From NASA Spacecraft these images of six of the planets in our solar system and the Earth's moon were taken. The picture mosaic consists of: Foreground the Earth rise over the lunar surface with the sun flare on the edge of the Earth's limb. First planet above the moon is Venus, top left to right are the planets Jupiter, Mercury, Mars and Saturn.

Picture Image taken by:

Earth	Apollo 17
Lunar Surface	Apollo 8
Sun	Apollo 12
Venus	Pioneer Venus
Jupiter	Voyager I
Mercury	Mariner 10
Saturn	Pioneer 11
Mars	Viking

FOR RELEASE:	December 14, 1979
PHOTO NO.	79-HC-455
	79-H-597

CHAPTER 7

LIGHT AND GRAVITY

There is a lot of evidence from the behavior of light that proves that gravity fields do indeed exist.

Light shifts to the red side of the spectrum in a star that is moving away from us.[15] This indicates that its waves are lengthened. As the light is coming towards us from a star that is leaving us and its waves are traveling through a less dense field as it goes through the wake of the star's field.

When a star is approaching us, its light is shifted to the blue side of the spectrum.[16] It's because the light is traveling through a gravity field that is being pushed ahead of the star and is denser and shortens the light's waves, changing its color.

The more it shifts in one direction or the other, the faster the star is moving towards us or away from us.[17]

Gravitational lensing is also proof of the gravity fields. It causes a single star to look like more than one. In some cases, one star looks like a pattern of four stars.[18] It occurs because the light from the star is being bent by the different densities of the gravity fields through which it is passing. It is bent just as it does when it enters water, a denser medium.

We know that light travels at the speed of light. But if light's electromagnetic system is so unique as to propel itself along forever at this speed and also maintain its same brightness, how can other particles that are not light travel at the same speed or even faster? They have no such system!

This question can't be answered until we accept the existence of the ether (the gravity fields).

Negatively charged neutrinos arrive at the Earth several hours before the light of the supernova which exploded thousands of light years away and produced them both at the same time.[19]

To travel that far, at that speed, a common subatomic particle has to catch a ride. It has to use light's system of transportation that can't be any other than the ether. Because, the ether, is the only thing which is everywhere in the Universe.

The neutrinos went faster because they have a small mass that was accelerated by the blast. If the neutrinos had no charge, they would have ended up as part of the gravity field and not have been propelled through it.

Light travels millions of times faster than the orbiting expanding gravity waves advance, but they are already everywhere for light simply to be propelled through their fields.

Light is the particle (photon) and gravity is the wave seen with light in a wave chamber. It's the only reason we can see the wave at all because it travels much slower than light. Gravity waves are the part of light that advances in waves like those made by dropping a stone into a pool of water.

Light also has the longitudinal wave of the photon particle that is propelled through the gravity waves. This is how a photon of light can be shot through one slot of a wave chamber and it acts like it also entered a second slot as well. The wave in the second slot is the gravity wave. And no one claims to have ever seen a gravity wave.

A photon hits the other side of the chamber as a particle and it was a particle when it was shot into the chamber. They are always particles.

Light needs to be propelled. There is no way that light can travel the distance from the beginning of time and arrive with the same brightness and speed that it had when it was first produced without a field to get its push from.

Light from our sun takes a little more than six minutes to travel the 93,000,000 miles to the earth. The sun's gravity field is what makes its' light work so well.

A test was made several years ago by sending a high energy ray of light towards Venus just as it rounded the sun in its orbit. Venus's position was known exactly when the ray bounced back from it.[20]

The results showed that the light had been slowed somewhere during its round trip. The time for the trip showed by its speed that Venus would have had to be 50 miles closer to us for the light to have returned in the correct time.

It was concluded that the strong gravity field of the sun had somehow slowed the light because it came so close to it when it passed by.

However, when the light passed by the side of the sun it was no longer crossing the gravity waves of the sun, it was tangent to them. It was not being propelled by them for a short ways in both directions proving that light needs the ether to be light.

LIGHT AND THE BLACK HOLE

At the center of some galaxies there are black holes that have a ring of light circling them.[21]

This light is regular light except that it can't leave radially as it does from a star. The gravity waves are so compact in the field the photons can't pass through them. The light leaves with the new waves and only orbits a ways and goes out. It shows the outline of the field by lighting it up.

The light does not go far because it is not crossing the field's waves and it dies. The ring is constantly being fed by new photons but none of them pass through the dense field.

The ring of light is at the equator of the black hole showing the strength of the field is there in both directions. The ring of light around a black hole provides a very revealing truth about black holes, the gravity fields and the way that light works and the way it doesn't work.

The reason that we see it at all is that it tries to leave with the new waves and it can't cross the waves ahead.

All of the preceding examples of the movement of light prove without a doubt that light is traveling through an ether. It proves that the gravity fields exist, what shape they have and what they consist of.

Light's propellant is gravity's ether, the ether that transports gravity's controlling force throughout the galaxies of the Universe. The ether is invisible but because of it we are able to be here and to be seen.

There is no way that we can see it or feel it. It's like a black hole in that we know its there because of what it is doing.

Just like a black hole is of the stars, so are the gravity fields which transport star light as everything is of the stars.

CHAPTER 8

THE PROVING OF THE ETHER

The sun orbits its galaxy in the sea of subatomic mass of the galaxy's ether. The waves are expanding outward keeping the sun outward in its orbit. When the planets are between the sun and the center of the galaxy, they are also held outward somewhat by the galaxy's advancing waves.

This holds the planets in closer to the sun when they are between the sun and the galaxy. This produces the elliptical orbits of the planets.

The closer they are to the sun, the faster they travel. It causes their orbits to advance a little bit each time, making their nearest point to the sun move ahead of where it was the time before.

This is what Einstein attributed to general relativity and showed its cause to be the taking of the path of least resistance. He stated that space was curved and that the stars and planets orbited in the paths of least resistance, which is all very correct but he didn't tell us how or why.

Isaac Newton could compute the results of the planets' orbits. His math was used to find the existence of other planets. But he didn't know what produced the force except that it behaved like all other force with a cause and effect having equal but opposite forces and reactions. Newton believed in an ether of some type but could not prove what it was.

In the late 1880's tests were run to determine if the ether existed. Light was sent outward in two directions at the same time to mirrors and returned to see if the ether had slowed it in either one direction or the other. It was tried in all directions and it had no effect on its time for the round trip in any direction.

This meant the end of the ether. What they determined was there was no ether or it was orbiting with the Earth and that made no sense to them at all. But the ether does orbit exactly with the earth as it carries it along and holds it in its orbit.

Newton needed to have some point in space that did not move to reference all of his computed motions as everything in space was moving. His work required absolute motion because it was founded on it.[22]

He thought maybe the ether did not move and that was the only thing that was at rest. But later the ether experiments unhinged his work. As there was no ether.[23] And they gave the Nobel Prize for proving that there was no ether even though there is one. They have since given the Nobel Prize for the proof of the existence of gravity waves. Gravity waves fill gravity fields and they are the ether. (See page 4.)

There is only one spot in the Universe where there is absolute rest and it's at the center of creation. It has never moved. Newton's work can be founded here. It stayed and everything else left.

All of the motions in our solar system are true from his work as well as all of the motion in the Universe. But, by studying the speed of the stars in other galaxies it has been determined that their outer stars are traveling as fast as the nearer orbiting stars towards the center. This seems to turn Newton's work upside down. But, our solar system is powered by just one star.

The galaxies have billions of stars at their centers and billions of stars in their orbits. All of their fields are eventually melded together. The outer stars are powered by all the other stars in the galaxy, thus enabling the outer stars to travel as fast as the inner stars.

To calculate their motion by Newton's methods all of the forces need to be accounted for. Newtonian physics abounds throughout the Universe just as it does here. For every action there is an equal but opposite reaction and $F = ma$ *(Newton)*. Amen

GRAVITY IS A PHYSICAL FORCE

Yes, there are many proofs of the gravity fields. They form the elusive ether that no one has ever been able to prove exists.

Scientists even have predicted that there is a gravity particle. They have named it a graviton. It is supposedly massless and has no charge.

Scientists also suspect that there are gravity fields.

Both Einstein and Newton have called gravity a force field.

There have been many tests conducted to prove the existence of the graviton. No one has ever proven their existence even though they were supposedly the most abundant particle.

When it was first devised, the inventors of the graviton thought that it needed to be a massless particle so it could travel at the speed of light relaying messages back and forth. There are no massless gravitons which can repulse and attract at the same time.

Gravity particles have mass and these particles are of several different types. The mass reacts to the forces placed on it and obeys the classical laws of physics that we know and understand.

There are many strange and wonderful things happening in the Universe, none of which can be explained by a force of gravity of attraction.

A third of the galaxies in the Universe have a disc shape like our own galaxy, the Milky Way.[24] Pictures of several galaxies show their spinning pinwheel rotations. They're shaped like a disc, thicker at the center and taper down, getting thinner toward their circumferences.

This shows their strength as being at their equators. Similar to the shape of an atom, our solar system and the rings of planets and the paths which satellites of the planets take. A true force of attraction would produce spherical galaxies only.

The strength comes from the length of radius throwing the field and size of the mass behind it. Many galaxies don't have a disc shape.

The question here is why don't they? Doesn't gravity work the same everywhere?

Yes, gravity does work the same everywhere. "The reason they are no longer in a disc shape and are round, oblong, or another shape is because they have gotten too close to each other."[25] Their fields have worked against each other. "Some galaxies have even passed through other galaxies and have combined with each other."[26]

When the Universe was created, the unbalanced matter forming the galaxies rotated in whichever direction it was unbalanced in. Therefore, some of the galaxies rotate clockwise and others counter-clockwise. "Some roll like a tire, others slant on a pinwheel angle."[27] Their positions are dictated by its matter's momentum, proving that each galaxy acts differently and is independent. They obey the forces acting on their own mass.

CHAPTER 9

THE BIG PICTURE

There is an ongoing proof that one rock has no attraction toward another. It is that the Universe continues to accelerate and separate in its expansion. The galaxies are diverging as they move outward.

There is a so-called critical mass needed to pull it back together, as though more mass is being developed and has an attraction for other mass. There is always less mass in the Universe. The stars generate heat. The heat part used to be matter and it's now heat energy. Mass is being used up. It is being used up because all of the mass that ever existed was produced before the Big Bang. No new mass is being made. The stars self-destructions are causing less mass in the Universe as their heat energy is produced.

There is also much more mass in the Universe than that which can be seen. It's in the form of Dark Matter and Brown Matter. Dark Matter consists of Black Holes and gravity fields of the stars and planets, all of which cannot be seen. The Brown Matter consists of objects which do not shine and cannot be located.

The separation of the Universe is occurring faster and faster as it accelerates. It is spreading apart and has been for over 10 billion years and will for another 10 billion years. If gravity were a force of attraction, it would have slowed down the runaway motion of itself long ago but it's going faster than ever. So it's not a force of attraction and it continues to act mechanically producing forces which create equal reactions to the expansion.

The fields of the galaxies control their own galaxies. Most of them don't even come into contact with one another. They diverge away from each other and tip in all different directions. They have many different shapes and sizes. We know that at least a third of them are in the shape of discs whose fields clearly show that they can't make contact with each other. It's like the planets whose fields all tip and no longer disturb each other because their fields no longer make contact.

We can only relax and enjoy the magnificent ride as we are propelled by the universal expansion. It's the only ride there is and its not coming back as it's going down a one-way street. Its gaining speed as it goes and has no way to slow itself down as gravity has no attraction between the galaxies because their fields don't surround one another.

We are located about two thirds of the way out from the center of our galaxy and our sun's motion is caused by a mass which is over two times as much as can be seen there. (See pg. 70) The total mass of our Galaxy then must consist of only about 3 or 4 time as much as can be accounted for and is not hundreds of times more as predicted by others.

The rest of the matter consists of mostly gravity fields, Black Holes near the center of the Galaxies, and Brown Matter which can't be seen as it is not a star but a moon or planet.

The mass of the gravity fields is being fed continuously by the destruction of the stars as it has been for over 10 billion years. New stars are continuously being formed from the gasses in the gravity fields and old stars are dying. The life and death of stars fill the fields of gravity with the invisible Dark Matter of the ether and Black Holes.

Black Holes have less matter than we thought because it's the acceleration of their mass that causes their strength, not just their mass alone. It's not an attraction of their mass which creates their force. It's the acceleration of it caused by their own super rotational speeds producing their huge gravitational force field.

In our Galaxy, mass is forced to swirl. We know this because the galaxy swirls around like a pinwheel. Every particle of mass, no matter how tiny, is given the force of its motion or its kinetic energy and because it all travels in curved paths, it is being normally accelerated. It's the gravitational normal acceleration, $a_n = \frac{v^2}{r}$ of the field's kinetic energy (its momentum) which is the fields gravitational acceleration.

The gravitational acceleration of the objects in the fields forces them to have a weight force. Our weight is a downward force as we are accelerated against the Earth just as the small balls of the Cavendish experiment were accelerated towards the larger ones.

METEOR TO A STAR TO A SUPER NOVA

I saw a story on TV of the meteor that exploded in Tungusbea, Russia in 1908. It produced a huge crater miles across. The meteor lit the night skies around the world. The meteor was huge and entered our atmosphere on a gradual incline towards the Earth as it circled it.

After over 90 years have passed, it now appears that the meteor exploded above the Earth without hitting it directly. It exploded like the atom bomb did over Hiroshima, only it was over 15 times larger than that atomic blast was.

The blasts produced the same butterfly pattern as the fallen trees laid identical to those at Hiroshima, including the circle of trees that remained standing but were without limbs. It was located near the center of the blast. The force there was directly down where the explosion was directly overhead and straight down.

The meteor's final approach to the Earth was determined to be about 30° and probably had gotten steeper in its flight.

The friction slowing its tremendous speed was from our atmosphere. It generated tremendous amounts of heat in the huge meteor. It became a bright star able to light the whole Earth as its protons were forced to fuse. It lit up the whole sky around the world, brighter than any distant Super Nova.

The meteor became a Super Nova in the white light of its nuclear explosion about six miles above the ground. It exploded just as stars do when they produce a nuclear Super Nova even similar to the Super Nova which produced the Universe. It was only a rock. It was composed of ordinary elements found in rocks, yet it became a bomb, a nuclear bomb.

There was no meteor found in the huge crater or below it or above it. There were only small pieces of broken rock that were determined to be part of the meteor.

In 1994 scientists observed a similar explosion of the meteor called Shoemaker that exploded at Jupiter. It as filmed. It produced an explosion on Jupiter making a crater the size of the Earth in Jupiter in very bright white light. That blast was over 1,000 times the size of the one in Hiroshima. There is every kind of energy in ordinary everyday matter. <u>Mass is energy and vice versa</u> as proven by Einstein's equation $e = mc^2$ with energy on one side of the equation and mass on the other. It can also be written as $m = \frac{e}{c^2}$.

Einstein studied the formula for kinetic energy for hundreds of hours before writing his energy equation for matter:

$KE = mv^2$ v is the velocity of the mass m

And

$E = mc^2$ c is the velocity of light and m is the mass of matter.

Light even acts as though it has mass. When a ray of sunlight hits a metal plate, the force of the light is able to knock electrons out of the plate when the rays hit it.

ABOUT THE PLANETS*

Planets	Specific Gravity	Accelerated Mass In Pounds	Mass in Equivalent Earth Mass	Radius In Miles	Average Distance From Sun In Millions Of Miles	Axis Tilt In Degrees
Mercury	5.42	6.6×10^{23}	0.05	1,509	36	0
Venus	5.25	112.5×10^{23}	0.85	3,750	67	-2
Earth	5.52	132×10^{23}	1	3,955	93	24
Mars	3.94	13.2×10^{23}	0.1	2,100	141	24
Jupiter	1.31	$41{,}976 \times 10^{23}$	318	44,250	485	3
Saturn	0.69	$12{,}540 \times 10^{23}$	95	37,320	886	29
Uranus	1.19	$1{,}941 \times 10^{23}$	14.5	16,260	1,800	98
Neptune	1.69	$2{,}244 \times 10^{23}$	17	15,390	2,000	29
Pluto	1.3±	2.6×10^{23}	0.02	1,000±	3,660	50±
Sun	1.4	$43{,}956 \times 10^{23}$	333,000	434,000	0	6
Moon	3.34	1.63×10^{23}	1/81	1,050	.240 from Earth	6.7
Io	3.55	1.97×10^{23}	1/67	1,158	.255 from Jupiter	0

*Most of the data for the planets is from two sources, Illustrated Atlas of the World, Rand McNally and Company 1985 and The New Solar System, Sky Publishing Corp. 1981. It has been converted to engineering units.

ABOUT THE PLANETS*

Planet	Average Orbital Velocity Miles/Hr	Orbital (Normal) Acceleration Miles/Hr^2	Time of Rotation Hours or Days	Present Rotation Velocity Mph	(Normal) Acceleration About Own Axis Miles/Hr^2	Days/Year
Mercury	107,000	318	58.6 Days	7.7	608	88
Venus	78,000	90	-243 Days	-4.0	1,518	224.7
Earth	66,700	49	24 Hrs.	1,000	1,590	364.25
Mars	53,700	20	24.6 Hrs.	537	850	687
Jupiter	29,300	1.72	9.864 Hrs.	28,367	18,100	4,333
Saturn	21,600	0.52	10.23 Hrs.	22,300	15,111	10,759
Uranus	15,200	0.13	15.5 Hrs.	8,500	6,518	30,685
Neptune	12,600	0.08	15.8 Hrs.	5,960	6,230	60,189
Pluto	10,700	0.03	6.4 Days	85	400	90,465
Sun	581,000±	Very Small	26.9 Days	4,167	176,000	200 Mil. Earth Years
Moon	2,300	22 About Earth	27.3 Days	10	425	27.3 About Earth
Io	37,700	5,574 About Jupiter	No Rotation	0	469	1.77 About Jupiter

CHAPTER 10

GRAVITY BY THE NUMBERS

The math and physics involved in proving that gravity is a mechanical force is wonderful.

Galileo showed us that a pendulums long swing takes the same amount of time as its short swing takes. This reveals an astonishing fact about how gravity works. Gravity has a built in time constant. Gravity's time constant produces all of the gravitational accelerations for the stars, planets and moons.

Make no mistake about it, gravity's time constant controls not only the pendulum swing, it controls the orbits of all satellites and their gravity forces. It does all of this with its natural built in time of orbit.

The time constant of gravity is revealed in Jupiter's rotational time. Jupiter rotates on its axis in 9.864 hours which is amazingly π^2 *hrs. (3.14 x 3.14 = 9.864 hrs).*

Jupiter rotates faster than all the other planets with Saturn being the next fastest at 10.23hrs per rotation. Uranus rotates in 15.5hrs and Neptune in 15.8hrs.

The rotation of all the planets other than Jupiter are all hampered by the constant action of the Suns field beyond them which holds them from escaping outward.

The inner most planets rotations have been slowed much more because their orbital acceleration are a lot bigger as they travel in shorter curved paths at faster speeds.

All the planets are driven to rotate by the constant rate of their orbiting fields at their surfaces. All but Jupiter are slowed by the Sun's constant force acting to hold them inward.

Because Jupiter is so massive, (318 times more mass than Earths), its rotation is not effected by the Suns action on its back side. Jupiter is located 485,000,000 miles from the Sun and gravity's strength is decreased by the distance squared d^2. The combined momentum of Jupiter's own rotation and its huge orbiting gravity field are able to overpower the Suns effect which would slow its rotation. Jupiter's size and location permits it to rotate at the natural rate that gravity produces. It rotates exactly with its field of gravity at its equator's surface in 9.864 hrs.

This is the time constant of gravity, 9.864 hrs/revolution of its field at each star, planet, moon and for our star, at the Suns equatorial surface. It produces our day because it is slowed by the Suns action which keeps us in orbit. It also produces our year as well as providing the gravitational acceleration for all heavenly bodies.

All fields of gravity orbit their sources in natural time. Our time, 24 hrs/day and natural times, 9.864 hrs/day produce Kepler's constant $\frac{9.864 hrs}{24 hrs} = .41$. Time = $.41\, r\sqrt{r}$ (page 3).

All fields of gravity orbit in natural time just as the pendulum swing is controlled by the natural time of the gravity field.

Jupiter's natural rotation is the fastest, sustainable rotation by gravity in the Universe. Super gravity of course rotates thousands of times faster but it's powered by the collapse of a star and its field. Its field makes 1000's of revolution/sec. due to the huge difference around the dwarf star which it fell into and Red Giant star it was orbiting as its velocity stays the same at first but slows as its mass is used up.

The normal acceleration of the Suns field at its equator determines the actual orbital accelerations of each planets path proving gravity's time is 9.864 hrs.

Further, the 9.864 hrs. natural time of orbit of each planets field at their own equator surface mathematically produces their individual gravitational accelerations proving how gravity is developed by their fields.

Kepler's ($K = \frac{t^2}{r^3}$) discovery that each planets year and their distance from the Sun proves that gravity has a time constant and that constant is truly 9.864hrs. It is the clock in Kepler's clockworks. Kepler's constant $K = .167$ is used to solve for *t* and $t = .41r\sqrt{r}$.

T's value *.41* converts our time to nature's time $\frac{9.864 hrs}{24 hrs} = .41$.

The planets gravity fields:

Earth
Circumference = distance traveled around the earth
Time Traveled = 9.864 hrs

π dia.
Cir = 3.14 x 7910 = 24,837 miles (distance traveled)

$V = \frac{Dist}{Time}$ $V = \frac{24{,}837\ miles}{9.864\ hours}$ = 2500 mph *Velocity of the gravity field at the Earth's equator.*

Now using Galileo's equation for gravitational acceleration:

$a_n = \frac{v^2}{r}$ $\frac{velocity\ of\ field}{radius\ of\ earth}$

$a_n = \frac{2500\ mph \times 2500 mph}{3955\ miles} = \underline{1590\ m/hr^2}$ *acceleration of earths field*

Mercury

$$C = \overset{\pi}{3.14} \times \overset{dia.}{3018 \text{ miles}} = 9476 \text{ miles}$$

$$v = \frac{9476\,miles}{9.864\,hrs} = 960 \text{ mph}$$

$$a_n = \frac{960\,x\,960}{1509\,radius} = 610 \text{ m/hr}^2$$ *Acceleration of Mercury's field*

Mars

$$C = \overset{\pi}{3.14} \times \overset{dia.}{4200 \text{ meters}} = 13{,}188 \text{ miles}$$

$$v = \frac{13{,}188\,miles}{9.864\,hrs} = 1335 \text{ mph}$$

$$a_n = \frac{1335\,x\,1335\,mph}{2100\,miles} = \underline{850 \text{ m/hr}^2}$$ *Acceleration of Mar's field.*

The Sun's acceleration can also be computed by using Galileo's equation for acceleration.

$$a = \frac{v^2}{r}$$

First, we need to find the circumference of the Sun. C = 3.14 x 868,000 miles, so C = 2,725,525 miles.

$$v = \frac{2{,}725{,}525\,miles}{9.864\,hrs} = 276{,}000 \text{ mph}$$

$$a = \frac{v^2}{r} = a = \frac{276{,}000\,mph\,x\,276{,}000\,mph}{434{,}000\,miles} = 176{,}000\ m/hr^2$$

The earth's acceleration about its own axis is 1,590 m/hr^2 compared to the Sun's 176,000 m/hr^2.

All the planet's natural surface velocities of their fields are found by dividing their circumference by time or 9.864 hrs. This is the natural velocity that gravity produces for all fields.

The earth also has an orbital acceleration around the Sun. Its acceleration in its orbit comes from acceleration of the Suns field about its equator.

Earth's velocity is known to be 66,700 *mph* around the Sun.

$$a = \frac{v^2}{r} \text{ (orbital) } \frac{66{,}700\,mph\,x\,66{,}700\,mph}{93{,}000{,}000\,miles} = 49\ m/hr^2$$

The earth and all the other planets get their orbital acceleration from the Sun's acceleration field.

PROVING THE SUN'S ACCELERATION AT THE PLANETS

The following exercise proves that the Sun's acceleration about its own axis fills its gravity field with its own acceleration. This acceleration is different at different distances from the Sun as it is reduced by d^2 at the distance *d*.

The Sun's acceleration of its field at 93,000,000 miles, Earth's distance, is its acceleration about its own axis reduced by the distance squared. The 9.864/24 = .41 conversion factor to our time from gravity's time is necessary.

$$\text{Sun's } a_n = \frac{176{,}000 m/hr^2}{.41 x 93 x 93} = 49 \, m/hr^2$$

which is the same as earth's orbital acceleration. Distance is in millions of miles

This is a very important proof.

First, it proves that the Sun's acceleration is correct at 176,000 m/hr^2 and, second, it proves the Sun's acceleration of its field at certain distances is exactly the acceleration given to anything at that certain distance.

This is why Kepler had so much success with his famous third law of motion for the planets.

The following are the accelerations for each planet in its orbit computed by using Galileo's formula

$$a = \frac{v^2}{r}$$

This is done by using the velocities of each planet in its orbit and the distance each is from the sun.

As will be shown, each is equal to the Sun's fields acceleration at each particular distance from the Sun. This provides a convincing proof that the Sun's normal acceleration is indeed 176,000 m/hr^2at its equator and that its acceleration towards itself is exactly reduced by .41 (*d* x *d*) in its field of gravity. *The Sun's gravity field is truly the Sun's own acceleration produced by gravity.* The acceleration is provided by gravity's field as it forces its mass to keep changing its direction.

Venus's orbital acceleration by the Sun's field:

$$a = \frac{78{,}180 \, mph \, x \, 78{,}180 \, mph}{67{,}000{,}000 \, miles} = 91 \, m/hr^2$$

Sun at Venus's orbit:

$$a_n = \frac{176{,}000 m/hr^2}{.41 x 67 x 67} = 95 \, m/hr^2$$

Mar's orbital acceleration by the Sun's field:

$$a = \frac{53{,}000 \, mph \, x \, 53{,}000 \, mph}{141{,}000{,}000 \, miles} = 20.7 \, m/hr^2$$

Sun at Mar's orbit:

$$a_n = \frac{176{,}000 m/hr^2}{.41 x 141 x 141} = 20.7\ m/hr^2$$

Jupiter:

$$a = \frac{29{,}130\,mph\,x\,29{,}130\,mph}{485{,}000{,}000\,miles} = 1.73\ m/hr^2$$

Sun at Jupiter's orbit:

$$a_n = \frac{176{,}000 m/hr^2}{.41 x 485 x 485} = 1.73\ m/hr^2$$

All the planets prove the same.

Mercury:

$$a = \frac{107{,}000\,mph\,x\,107{,}000\,mph}{36{,}000{,}000\,miles} = 318 m/hr^2$$

Sun at Mercury's orbit:

$$a_n = \frac{176{,}000 m/hr^2}{.41 x 36 x 36} = 330\ m/hr^2$$

This is the proof of gravity's time constant. The Sun's field acceleration is computed using 9.864 *hrs* and the actual acceleration of the planets orbits are proven to be developed by it.

GRAVITY'S CAUSE BY MECHANICS IS REVEALED IN THIS EQUATION FOR THE ACCELERATION OF GRAVITY

$$g\ ft/sec^2 = \frac{a_n\ x\ sp.g}{274 miles / hr^2}$$

Earth $$g = \frac{1590 m / hr^2\ x 5.52}{274} = 32.2\ ft/sec^2$$

Moon $$g = \frac{425 m / hr^2\ x 3.34}{274} = 5.18\ ft/sec^2$$

Mars $$g = \frac{850\ x 3.94}{274} = 12.2\ ft/sec^2$$

Venus $$g = \frac{1518 x 5.25}{274} = 29.08 ft/sec^2$$

Mercury $$g = \frac{608 m / hr^2\ x 5.42}{274} = 12\ \ ft/sec^2$$

The product of *sp.g* and a_n are directly proportional to their value of gravitational acceleration. The 274 constant converts units to ft/sec^2. (.000407 x 8.99) = .00365 and 1/274 = .00365 (See pg. 59)

<u>Each planet's acceleration of gravity can be determined from its fields normal acceleration of its specific gravity proving its mechanical origin!</u>

The normal acceleration (a_n) of each planet has been computed by using the <u>velocity of its field based on its 9.864 hour orbit which leads to each of their gravitational accelerations,</u> proving their fields all orbit them in 9.864 *hr* at their equators.

The a_n is converted to ft/sec^2 by

$$\frac{5280\ ft / mile}{3600 \sec / hr} = 1.46\ ft/sec \times 1.46\ ft/sec = 2.15\ ft^2/sec^2$$

$$\frac{(1.46)^2}{5280 ft / mile} = \frac{2.15 ft^2 / \sec^2}{5280 ft}\ .000407\ ft/sec^2$$

The conversion of the planets a_n from *miles/hr²* to *ft/sec²*.

Earth	.000407 x 1590 m/hr^2	= .648 ft/sec^2
Mars	.000407 x 850 m/hr^2	= .346 ft/sec^2
Venus	.000407 x 1518 m/hr^2	= .618 ft/sec^2
Mercury	.000407 x 608 m/hr^2	= .247 ft/sec^2
The Moon	.000407 x 425 m/hr^2	= .172 ft/sec^2

The kinetic energy of a gravitationally rotated sphere can be expressed in units form.

$KE = mv^2$ $\quad$ $m = (sp.g \times vol.) = mass$

$KE = (sp.g. \times vol.)v^2$

The earth has a specific gravity of 5.52 or 5.52 times as dense as water which is 1.

The formula for the volume of a sphere is $4/3\pi r^3$. Using 1 for the radius, this becomes $1.33 \times 3.14 \times 1^3 = 4.19$ which is the unit volume of a sphere with a radius of 1.

The velocity of 1 squared is also 1. The unit of velocity squared must be multiplied by 2.15 as it has been for the a_n conversion in the acceleration portion of $g = a_n \times KE$. (See pg. 58)

The *KE* of earth's rotation in units is:

$$KE = \underset{sp.g}{5.52} \times \overset{8.99}{(\underset{vol}{(4.19)} \times \underset{V^2}{2.15})}$$

Mass above 5.52 x (4.19) x 2.15); *sp.g .. vol .. V²* below.

$KE = 49.73$ *units*

$g = 49.73 \times a_n$

$g = 49.73 \times .648\ ft/sec^2 = 32.2\ ft/sec^2$

also

$KE = 5.52 \quad (8.99) = 49.73$

$4.19 \times 2.15 = 8.99$

	$g = (a_n\ ft/sec^2 \times sp.g)$ 8.99 = *grav. accel*		*
Earth	$g =$ (.648 x 5.52)	8.99 = 32.2 ft/sec^2	32.2 ft/sec^2
Moon	$g =$ (.152 x 3.34)	8.99 = 5.18 ft/sec^2	5.31 ft/sec^2
Mercury	$g =$ (.247 x 5.42)	8.99 = 12.03 ft/sec^2	12.4 ft/sec^2
Venus	$g =$ (.618 x 5.25)	8.99 = 29.17 ft/sec^2	28.9 ft/sec^2
Mars	$g =$ (.346 x 3.94)	8.99 = 12.26 ft/sec^2	12.2 ft/sec^2

* This column is a comparison, as it is the average values of g for each planet computed from their action with the Sun rather than from their own properties. These values are from the book The New Solar System by Beatty, O'Leary & Charkin. The units have been converted from m/sec^2 to ft/sec^2.

Specific gravity of each planet or star is the direct result of their fields gravitational acceleration. Therefore, their gravitational accelerations can be computed by using their specific gravities.
$g = a_n \times KE$ produces the gravitational acceleration of all the heavenly bodies.

This phenomenon I call a Dave in honor of my brother.

The calculation of the acceleration of gravity at the surface of each planet starts with its own radius.

The larger planets' radii have been deformed due to their own centrifugal forces.

To get the true value of their normal acceleration of gravity, it is necessary to correct for this ellipticity as their poles have been flattened and their equators are bulging.

To use their bulging radii inflates their true value of their acceleration of gravity. But in reality they decrease their a_n making it necessary to make a double correction for ellipticity.

Actually, this is really the convincing proof of what causes the force of gravity. Because by using their expanded radii, it gives a much greater value for their acceleration of gravity than can be computed from their actions towards the Sun.

Their computed acceleration of gravity from their own properties must agree with what is taking place between them.

The ellipticities for the planets are from the book The New Solar System by Beatty, O'Leary & Charkin, pg 219. Their values are computed by using this formula:

$$\frac{(R)equiter - R\ planetary}{(R)\ planetary} = ellipticity$$

The amount of the correction required is two times the planet's ellipticity as follows. It needs to be two times because the bulging decreases its acceleration. By using its bigger radius, it computes a larger acceleration. The first of two times just gets back to where it can be corrected. Because it is no longer a sphere, it decreases its a_n even though it has a larger radius.

Jupiter	.0637 x 2 = - .127
Saturn	.102 x 2 = - .204
Uranus	.024 x 2 = - .048
Neptune	.0266 x 2 = - .0532

These corrections are then multiplied by the *g* produced by using the enlarged radii getting the amounts to be deducted from the exaggerated values of *g*.

$g = (a_n\ ft/sec^2 \times sp.g.)\ 8.99$ *Inflated Values*

Jupiter	$g =$	(7.29 x 1.30)	8.99 =	85.2 ft/sec^2
Saturn	$g =$	(6.15 x 0.69)	8.99 =	38.2 ft/sec^2
Uranus	$g =$	(2.65 x 1.19)	8.99 =	28.4 ft/sec^2
Neptune	$g =$	(2.54 x 1.69)	8.99 =	38.6 ft/sec^2

	Corrections	Computed Values
Jupiter	g = (85.2 ft/sec^2 x (-.127) = (-10.8 ft/sec^2)	85.2 – 10.8 = 74.4 ft/sec^2
Saturn	g = (38.2 ft/sec^2 x (-.204) = (-7.8 ft/sec^2)	38.2 – 7.8 = 30.4 ft/sec^2
Uranus	g = (28.4 ft/sec^2 x (-.048) = (-1.4 ft/sec^2)	28.4 - 1.4 = 27 ft/sec^2
Neptune	g = (386 ft/sec^2 x (-.053) = (-2.1 ft/sec^2)	38.6 – 2.0 = 36.6 ft/sec^2

Comparisons

	g computed with owned properties and corrected for ellipticity	*g* converted from the *g* found in book New Solar System from the sun's action upon them.
Jupiter	g = 74.4 ft/sec^2	75.0 ft/sec^2
Saturn	g = 30.4 ft/sec^2	29.8 ft/sec^2
Uranus	g = 26.9 ft/sec^2	25.5 ft/sec^2
Neptune	g = 36.6 ft/sec^2	36.1 ft/sec^2

Clearly, the preceding shows the smaller planets are more nearly perfect spheres and their properties produce almost exactly the same value of *g* as has been computed by others using their action(s) towards the Sun.

Getting the inflated values of *g* for the larger planets results in using their enlarged radii. Their values of *g* truly would need to be corrected as they are not in the true shape of a sphere.

The planets' corrected radii, specific gravity, and 9.864 hour time of rotation for their fields combine to prove that gravity is indeed due to the mechanics of orbiting mass.

Using their distance to the sun and computing their gravity forces as Newton prescribed compares near to perfection of the values computed due to their own properties and mechanics.

It's amazing to me that some of the planets are of gas and Neptune is even a liquid, yet, they all form spheres. The Sun is a gas with a *sp.g* of 1.4. Jupiter has a *sp.g* of 1.31 which is a less dense gas than the Sun. Saturn is a very light gas, having a density of only .69. Neptune's *sp.g* is 1.69 and is a blue ball of liquids.

The thing that they all have in common is that they all have their own gravity fields. Their gravity field forces each of them into their spherical shape. The larger ones centrifugal force of rotation at their equator produces their midriff bulges, clearly showing they would fly apart if they were not held inward by the crushing force of their fields.

THE TRUE GRAVITY CONSTANT

The force of gravity consists of two forces: One, the fields rotating centrifugally outward of its mass pushes satellites outward and around. And two, its inward acceleration of mass which holds the satellites inward from escaping the orbits.

The two forces are equal but opposite at any point in a gravity field.

Isaac Newton's equation $F = \frac{m_1 m_2 G}{d^2}$ for gravity uses both m_1 and m_2, the mass of the source and the mass of the satellite.

Because the centrifugal force and the gravity force are equal for all satellites, their equations can be set equal, eliminating the sources mass.

Centrifugal Force Outward		Gravity Force Inward
$F_c = \frac{mv^2}{Gd}$	$=$	$F_g = \frac{mG}{d^2}$

Then the true gravity constant <u>G</u> can be found by solving for <u>G</u> by cross multiplying.

$$\frac{mv^2}{Gd} = \frac{mG}{d^2}$$

$$v^2 = \frac{G^2}{d}$$

(and taking the square root)

$$dv^2 = G^2$$

$$\boxed{G = v\sqrt{d}}$$

G is the True Gravity Constant

Then by using each planets velocity in *mph* for *v* and the square root of their distance to the Sun in miles, each one produces the same amount for *G* = 642,000,000, the true gravity constant.

$$\boxed{G = v\sqrt{d} \text{ also } G = d\sqrt{a}}$$

Every planet of the Sun reveals the gravity constant of $G = v\sqrt{d}$ of the sun to be 642,000,000 *mph* ± .0047.

Mercury	$G =$ 107,000 *mph* x	6,000 =	642,000,000
Venus	G = 78,200 mph x	8,190 =	641,000,000
Earth	$G =$ 66,700 *mph* x	9,645 =	643,000,000
Mars	$G =$ 53,700 *mph* x	12,000 =	641,000,000
Jupiter	$G =$ 29,332 *mph* x	22,000 =	645,000,000
Saturn	$G =$ 21,630 *mph* x	24,300 =	642,000,000
Uranus	$G =$ 15,184 *mph* x	42,460 =	643,000,000
Neptune	$G =$ 12,141 *mph* x	53,780 =	640,000,000
Pluto	$G =$ 10,700 *mph* x	60,000 =	642,000,000

The true gravity constant is 642,000,000±. For it to be off by 3 in 642 is 3/642 = .0047 x 100 = .47%. Less than ½ of 1%.

And the truth of the matter is if my distances and velocity were absolutely correct, the exact $G = v\sqrt{d}$ would all be exactly equal at an odd number around 642,000,000±.

By setting the acceleration equation and the gravity equation equal, we can solve for the real gravity constant. The results are almost flabbergasting.

Every planet in the solar system produces the same value of *G* from the Sun's field at their location in it. $G = v\sqrt{d}$.

The planet's actual velocity times the $\sqrt{\ }$ of the distance it is from the Sun gives a value of 642,000,000 *mph*.

The Sun's $G = v\sqrt{d}$, $G =$ 642,000,000 *mph* shall be called a David in honor of my father.

The true gravity constant has been obscured by an incomplete understanding of how gravity works. To me it's the wind beneath the wings of creation. Imagine how astonishing this is. Every planet's velocity times the square root of its distance from the Sun producing the same constant.

This development is quite exciting to me.

Imagine all the planets producing the same $G = v\sqrt{d}$ from Mercury to Pluto.

	Actual Velocity		Distance	
Mercury	107,000	x	6,000 $\sqrt{36{,}000{,}000}$	= 642,000,000
Pluto	10,700	x	60,000 $\sqrt{3{,}600{,}000{,}000}$	= 642,000,000

In the process of taking the square root of a quantity, its unit is dropped because there is no unit for the square root of a mile. This number is then multiplied by *mph*, the velocity giving 642,000,000 mph.

By using the gravity constant in the gravity equation and the centrifugal force or acceleration equation, the two forces can be computed for each planet and it is shown that they are equal.

Gravity Inward toward Sun	Centrifugal (acceleration) Outward away from Sun
$F = \frac{mG}{d^2}$	$F = \frac{mv^2}{Gd}$

Mercury's force toward the Sun. — Mercury's centrifugal force in orbit.

$$F = \frac{.05^* x 132 x 10^{23} x 642{,}000{,}000}{36{,}000{,}000 x 36{,}000{,}000} \qquad F = \frac{.05^* x 132 x 10^{23} x 107{,}000^2}{642{,}000{,}000 x 36{,}000{,}000}$$

$F = 320 \times 10^{15}$ *m lbs/hr* = $F = 320 \times 10^{15}$ *m lbs/hr*

Venus

$$F = \frac{.81^* x 132 x 10^{23} x 642{,}000{,}000}{67{,}000{,}000 x 67{,}000{,}000} \qquad F = \frac{.81^* x 132 x 10^{23} x 78{,}167^2}{642{,}000{,}000 x 67{,}000{,}000}$$

$F = 1{,}524 \times 10^{15}$ *m lbs/hr* = $F = 1{,}524 \times 10^{15}$ *m lbs/hr*

*Mercury's mass is .05 times Earths and Venus's mass is .81 of Earths.

By removing the mass of the Sun from Newton's equation, it allows the force acting on the planets to reveal the true value of the gravity constant. It's like removing the mass of the Earth from the process of weighing yourself. You don't need it to get the answer, and it is less complicated and more understandable.

EACH BODY HAS ITS OWN VALUE FOR G

All the planets have their own value of *G*. It can be determined by the velocity of each of their satellites times the square root of the distance to the center of the planet.

Earth's *G* $G = v\sqrt{d}$

From the moon's motion

$$G = 2300 \overset{490}{\sqrt{240{,}000}}$$

$$G = 1{,}127{,}000 \; mph$$

Jupiter's *G*
From Io's motion

$$G = 37{,}700 \overset{505}{\sqrt{255{,}000}}$$

$$G = 19{,}038{,}000 \; mph$$

Uranus's *G*
From its motion
Of its moon Oberon

$$G = v\sqrt{d}$$

$$G = 7{,}060 \overset{602}{\sqrt{363{,}320}}$$

$$G = 4{,}250{,}120 \; mph$$

Neptune's *G*
From its motion
Of its moon Triton

$$G = v\sqrt{d}$$

$$G = 9{,}803 \overset{470}{\sqrt{220{,}600}}$$

$$G = 4{,}597{,}650 \; mph$$

The planets moon's velocities and distances are found in the book: The New Solar System by Sky Publishing Corp. 1981

CHAPTER 11

AS MERCURY GOES, SO GOES THE GALAXY

The distance a moon is from the planet is determined by the moon. It would be at the exact same distance away from the Earth if it were in its orbit as it would be from Saturn, Mars or Jupiter. The size and mass of a moon is balanced in an orbit at a certain distance from a source. Because the field's force is balanced both outward and inward for their size there. This same size and mass finds its place at exactly the same distance even though it might be at a larger planet because the outward force and inward forces are both bigger but still are equal and the same moon is held at the same distance.

Just look at our moon and Io. They are almost the same distance from the Earth and Jupiter. They are almost the same size and mass.

By using any moon's distance in Mercury's orbit, a velocity can be computed mathematically for it or anything at that distance using Mercury's *G*. Then, by comparing that moon's planet's *G* to Mercury's, its actual velocity can be determined.

This works even for the Sun and its orbit of the galaxy.

Uranus's satellite Oberon has an actual velocity of 7,060 *mph*. It is 363,320 miles from its planet Uranus.

Using the $\sqrt{d}$ for Oberon at Mercury. $G = v\sqrt{d}$.

Mercury's $G = 240{,}000$* (as Mercury has no moon).

*Mercury's *G* has been calculated by $\sqrt{\Delta m} = \Delta G$ by comparing it to Earth and Jupiter. (See pg 68)

$$240{,}000 = v\overset{602.8}{\sqrt{363{,}320}}$$

$$\frac{240{,}000}{602.8} = v = 398.14\ mph$$

It would have a theoretical velocity in Mercury's orbit of 398.14 *mph*.

$$\frac{Uranus\,G}{Mercury\,G} = \Delta G = \frac{4{,}220{,}000}{240{,}000} = 17.58$$

398.14 *mph* x 17.58 = 7060*mph* is the velocity computed from Mercury's orbit which agrees with the actual velocity.

Neptune has a moon, Triton, that is located 220,600 miles from Neptune and it travels at an average speed of 9,803 *mph*.

Using its distance at Mercury, we have $G = v\sqrt{d}$.

$$240{,}000 = v\overset{469}{\sqrt{220{,}100}}$$

$$\frac{240{,}000}{469} = v = 511mph$$

$$\Delta G = \frac{Neptune}{Mercury} = \frac{4{,}597{,}650}{240{,}000} = 19.157$$

$$511 \quad x\ 19.157 = 9803mph$$

Triton's velocity computed by using its distance in Mercury's
$G = v\sqrt{d}$ then multiplied by the ΔG of the two hosts giving its actual velocity at its present location.

It is now necessary to understand that $\sqrt{\Delta m} = \Delta G$.

The difference in mass of Mercury and Jupiter is 318 x 20 = 6,360 times.

All the planets in this book are compared to Earth's mass. Jupiter's mass is 318 times larger than Earth and Mercury is 20 times less than the Earth. Jupiter's mass is then 6,360 times larger than Mercury.

$$\sqrt{\Delta m} = \sqrt{6{,}360} = 79.7$$

Then:

$$\Delta G \frac{Jupiter}{Mercury} = \frac{19{,}038{,}000}{240{,}000} = 79.3$$

Proving that $\sqrt{\Delta m} = \Delta G$.

Believe me, this is the same number. We are working with two different planets, Jupiter being the largest and Mercury the next to smallest one.

This relationship is true for all the heavenly bodies.

All the planet's *G*'s can be calculated the same way as the Sun's has been. The Sun's satellites all produce $G = v\sqrt{d}$ for that particular planet just as Earth's $v\sqrt{d}$ in the Sun's orbit which reveals the value of the Sun's *G* or its comparative strength of 642,000,000.

The Sun's velocity of 581,000 *mph* can even be computed using Mercury's orbit and then comparing it to the galaxy's *G*.

First it is necessary to determine the value of G for the hub of the galaxy.

Using the Sun's orbit of the galaxy of 581,000 *mph* at 162 x 10^{15} *miles from its center*:

$$G = 581{,}000\sqrt{162\,x\,10^{15}}$$

$$G = 581{,}000 \times 402{,}490{,}000$$

$G = 233.8 \times 10^{12}$ for the galaxy out to the Sun from the Sun's distance and velocity.

Then:

$$\Delta G = \frac{Galaxy}{Mercury} = \frac{233.8\,x\,10^{12}}{240\,x\,10^{3}} = \Delta G = 9.74 \times 10^{8}$$

Using the velocity computed at Mercury's orbit from the Sun's distance to the center of the Galaxy, one is able to compute the Sun's own velocity around the Galaxy.

$$G_m = v\sqrt{d}$$

$$240{,}000 = v\sqrt{162\,x\,10^{15}}$$

$v = .0005962$ *mph* (Theortical velocity in Mercury's orbit.)

The Sun's computed velocity from Mercury's orbit

$v = (v)$ at Mercury x ΔG

$v = .0005962 \times 9.74 \times 10^{8}$

$v = 580{,}700$ *mph* computed from Mercury's orbit

The Sun's average $v = 581{,}000 \pm$ *mph*

Now by knowing that $\sqrt{\Delta m} = \Delta G$ it is possible to compute the mass of the hub of the galaxy.

$$\frac{G \quad Galaxy's\,Hub}{G \quad Mercury} \quad \Delta G = 9.74 \times 10^{8} = \sqrt{\Delta m}$$

$$\sqrt{\Delta m} = 9.74 \times 10^{8}$$
$$\Delta m = (9.74 \times 10^{8})^{2}$$
$$\Delta m = 94.87 \times 10^{16}$$

Mercury's mass times Δm

$m = 94.87 \times 10^{16} \times 6.6 \times 10^{23}$

Mass of the Galaxy responsible for the sun's velocity.

$$m = 62.6 \times 10^{40} lbs.$$

This is the mass of 142 billions of our Sun which is over twice what the whole visible Galaxy was once thought to have.

This is the mass of the Galaxy which controls the orbit of our Sun. It includes the mass of all that exists from the center of the Galaxy out to our Sun.

It's the mass of everything that produces the Sun's orbit. A very large percentage of it is of the invisible gravity fields of all of the stars whose fields combine to orbit the sun around the Galaxy. It also includes brown matter and black holes which are not seen.

By being able to compute the Sun's actual velocity using Mercury's orbit is the proof that Newton's classical physics applies everywhere.

All of the stars and their gravity fields' mass inside of the Sun's orbit contribute to its orbital velocity.

The amount of the mass causing the orbit of the Sun is able to be computed by comparing the different values of *G* for Mercury and the Galaxy. Using the Sun's orbit $G = v\sqrt{d}$ which is known

$$\frac{Galaxy\, G}{Mercury\, G} = \Delta G$$

Then: $$\sqrt{\Delta m} = \Delta G$$

$$\Delta m = G^2$$

The mass of any galaxy can be computed by knowing its $G = v\sqrt{d}$ and $\sqrt{\Delta m} = \Delta G$.

All of this proves that Newton got it right. For every action there is an equal and opposite reaction.

The center of the source's field of gravity is the fixed location necessary to compute the forces. Even though it is moving straight outward, it is rotating on its axis. The center of rotation is fixed with respect to its field which is also moving straight outward with it.

GALILEO

Galileo devoted his life to working with the mysteries of gravity. He was persecuted, beaten, and then placed under house arrest for making his work known. It was thought that studying such things as heavenly bodies was sacrilegious in those early days of the 1500s.

He is a giant to the understanding of gravity's operation. He developed the fundamental equation for its normal acceleration $a_n = \frac{v^2}{r}$. He dropped things of different mass from the top of the Leaning Tower of Pisa and determined that light objects fell at the exact same rate as heavy objects do. Everything is being accelerated at the same rate and that rate is 32.2*ft/sec*^{2}.

Galileo discovered the pendulum effect. The pendulum is a gravity device. He determined that the pendulum takes the same amount of time to make a short swing as it does to make a long swing. This is the device used to make the first clock. As its swing slows down it is still taking the same amount of time as the longer one. It's so perfect that time can be measured precisely with it.

<u>By knowing that this is true of gravity, it's possible to prove how gravity works. It's a wonderful discovery.</u>

The same gravity field that controls the pendulum swing also controls the orbits and rotation of the stars and planets. The pendulum makes a short and long swing in the same amount of time and a planet's field travels around a large planet in the same amount of time as it does a small one because of this same action of the field. This amount of time is exactly 9.864 *hours*. This is the rate which gives the Sun's field its force. It is also the rate which gives Kepler's discovery its constant of $\frac{9.864 hrs}{24 hrs} = .41$. (See pg 3) It converts gravity's time to our time. 9.864 hours is also the rate which gives gravity its acceleration towards the planets and stars. Each one of them has a field of gravity which circles it in 9.864 *hours* at their equator's surface and can be used to calculate their gravitational accelerations.

Jupiter rotates in 9.864 *hours* also which is 3.14 x 3.14 = 9.864 or π^2*hrs*. For it to do this its surface velocity divided into its diameter needs to be equal to 3.14 and it does. Jupiter rotates at the fastest natural rotation that gravity can produce. Super gravity rotates many times faster but its cause is not from a natural gravity field but from a huge collapsed star which forms a Dwarf Star. The normal acceleration of all the planet's fields are calculated from $a_n = \frac{v^2}{r}$ by using the speed of their gravity fields from the time of their orbit of 9.864 *hrs*. $C = \pi d = Dist.$ and $\frac{Dist.}{Time} = Velocity$ and $\frac{d\pi}{v} = 9.864$ and $\frac{dia.}{v} = 3.14$.

By using 9.864 *hours*, the velocity of the Sun's field at its surface can be determined and its normal acceleration of its gravity field is computed from $a_n = \frac{v^2}{r}$.

The fact that the gravitational time of 9.864 *hours* can be used to calculate the gravitational acceleration of the Heavenly Bodies is wonderful.

I call this phenomenon Lucy in honor of my mother.

One revolution of the Sun's field at its surface in 9.864 *hours* produces time in our solar system. Using Kepler's third law of motion (see pg 3), the number of days in each planet's year can be calculated by knowing only how far the planet is from the Sun. Gravity's time of 9.864 *hours* can be divided by our 24 hour day to produce the same constant as Kepler's .41.

Jupiter's rotation of 9.864 *hours* unlocks the mysteries of gravity. It's flabbergasting that Jupiter rotates in π^2 hours. Jupiter, combined with its field, is so massive that Jupiter's constant rotation is sustained at the maximum rate gravity can naturally produce. Jupiter is 485,000,000 miles from the Sun and the Sun is able to control Jupiter's orbit, but it has no affect on slowing its natural rotation as it continues to rotate exactly in step with its field.

CHAPTER 12

GRAVITY FIELDS

Jupiter is by far the largest planet of the Sun. It contains 318 times the mass of the Earth yet it is able to rotate on its axis faster than Earth or any of the other planets. It rotates in 9.864*hrs* compared to Earth's 24*hr* rotation.

Gravity fields orbit their sources and produce each of the planets rotations. Gravity fields were once thought to be massless and to be located in the vacuum of space traveling at the speed of light. However, it has now been proven that the so-called vacuum of space which the Universe occupies is a teaming sea of subatomic particles.

These particles are released from each star at millions of tons/sec. They form the invisible, undetectable, seas of the gravity fields. Subatomic particles are the remains of atoms whose protons have fused together in the nuclear reactions of the stars. They form seas of gravity fields thousands of times thinner than air. The particles have mass retaining that part of matter not converted into heat and light energy by the fusion process.

The Universe has recently been proven to be flat and to have been developed at one location. It is also expanding with acceleration into a larger disc shape.

In order for the Universe to have developed at one location and to be flat it is necessary for it to be falling also. Because if it were not falling it would be spherical.

The Big Bang which produced the Universe had to be caused by an equally inward force from all directions towards its center in order for it to have taken place. The huge nuclear explosion resulting would have been equally outward in all directions forming a spherical shape.

The Universe expands in a flat shape growing into the void of nothingness, filling it with what there is. There is no friction to ever slow its expansion.

The Universe falls through the void at the same time. The upward acceleration of the Big Bang is overcome by the downward acceleration of the falling and the downward acceleration of the Big Bang is caught up with it by the falling creating a flat Universe.

The Universe does not rotate only its pieces do. The Universe has no gravity field only its pieces do. The Galaxies, the stars, planets and moons all have gravity fields. The Universe expands straight outward with its pieces orbiting within their galaxies.

The gravity fields contain over twice as much mass as can be seen in the Universe. The forces on the orbiting subatomic mass within the gravity fields powers the orbits of all satellites, all stars, all planets and moons.

All the mass of the heavenly bodies and their gravity fields of subatomic particles have kinetic energy due to the motions given to them by the expanding and falling Universe.

Stars are spewing the remaining parts of their atoms into their own surroundings and are enveloped by them forming the seas of their gravity fields. The planets and moons recycle the subatomic seas of their sources field creating their own gravity fields of them.

Gravity fields orbit their source, powered by the kinetic energy of their inherent motions given to them by the Universe.

Any mass acted on by two forces from two different directions must swirl. The swirling produces the orbits of the mass in the gravity fields which produces the rotations of their sources. Orbiting particles of gravity are forced to travel around their sources in curved paths.

Subatomic particles are being centrifugally thrown from their inner orbits and are forced into larger, slower moving orbits further from the source. They keep progressing around and outward taking larger orbits as the new material behind them forces them outward taking their place. Eventually they make their way into the far reaches of the orbiting gravity fields.

Fields of gravity are mechanically operated by the physical force placed on them. Their momentum, given to them by the expanding and falling Universe, forces them to swirl and orbit. It causes the stars and planets to act like toy tops with a never ending falling plunger which endlessly provides the spinning and creates the centrifugal force as it turns.

At any location in a gravity field there is an inward force and an outward force. The fields orbit in curved paths and they expand centrifugally outward. This keeps satellites outward in their orbits and the curved orbiting field beyond the satellite crowds them around and inward holding them from leaving their orbits. The satellites travel at the speed of the field at their location in it.

Anything which travels in a curved path is being accelerate inward 90° from every point on that path. This is called normal acceleration and its amount is computed by $a_n = \frac{v^2}{r}$. In words the velocity of the satellite in *mph* is squared and then divided by the distance it is from the center of the Sun (for a planet). This is also true for each moon of the planets by using their speeds and distances to the center of their planets.

All the fields of gravity push centrifugally outward and crowd around and inward in orbit. Depending on which side of a wave a satellite is located on in orbit these forces either push on it outward or hold it inward by the same waves. When an object is in orbit the outward force acting on it is exactly equal to the inward forces acting on it at its location. Satellites stay in orbit at the location in the field where the two forces are balanced for their size and mass and are carried along at the speed of the field at that location.

The individual subatomic particles are so tiny and petite that their presence is neither seen nor felt by us as they whiz by. Yet their inward acceleration collectively adds together from the outer reaches of the gravity field to the surface of the Earth to produce our acceleration of gravity of 32.2ft/sec^2.

Galileo discovered that a pendulum (a gravity device) takes the same amount of time for it to make a long swing as it takes for it to make a short swing. The cause of this important discovery is the existence of the gravity field and its behavior.

As it turns out this phenomenon is the key to proving that gravity fields exist and that they operate in the manor thus far described.

Because the field causes the swing time to be the same for all lengths of swings it reveals how the fields work. Fields of gravity orbit large spheres in the same amount of time that they do smaller spheres. This is the basis used to prove that gravity fields exist and how they produce the gravitational acceleration for each of the heavenly bodies. It mathematically proves how the speed of the Sun's field at its equator produces the orbital speed of each of the planets and how the normal acceleration of the Suns field at its equator develops the orbital normal acceleration of each planet.

The fact that Jupiter is the largest planet and rotates faster than all the others is very suspicious. It rotates in 9.864*hrs* exactly in time with its own gravity field which produces its rotation. The fields of all the other planets orbit their planets in 9.864*hrs* also, but each of their rotations have been hampered in various amounts by the action of the Sun's field on them from behind as it holds them inward in their orbits.

Jupiter rotates in 9.864*hrs* while Saturn takes 10.23*hrs* to turn around. Even though Saturn is nearly twice as far from the Sun as Jupiter its size and mass are less than a third of Jupiter's. Saturn's rotation is being only slightly retarded by the Suns effect on it from behind it.

Jupiter, with its 318 Earth masses, is so massive and is located 485,000,000 miles from the Sun so that the Suns action holding it from leaving its orbit from behind is over powered by its own natural rotation caused by its own gravity field's force. The rotating mass of Jupiter and its massive field are forced to turn at 9.864*hrs* by the momentum given to them by the expanding and falling Universe. Jupiter over powers the Suns retarding rotation force which slows all of the other planets rotations.

Uranus rotates in 15.5*hrs.* and Neptune in 15.8*hrs*, both over 4 times further from the Sun than Jupiter but they are tiny in comparison to it.

The rotations of the smaller inner planets to the Sun are very much effected by the Sun's field force on their back sides. They travel much faster than the outer planets and move in smaller curved paths causing them to develop a huge centrifugal force which pushes them out against the Sun's field which is pushing inward holding them from leaving their orbits. This action has almost stopped Mercury from rotating. Mercury rotates only once in 58.6 *days* because of it.

Venus even rotates backwards because its rotation is so hampered by the Sun's field that the Earth's field caused Venus to rotate backwards. The Earth's field engages Venus similar to how gears work causing the gear which is being driven by the other to rotate in the opposite direction. Earth's field causes Venus to make one rotation backwards every 243 *days.*

Earth and Mars are further away from the Sun than Mercury and Venus are and they both rotate on their axis in about 24*hrs* each.

All of the gravity fields orbit their sources in 9.864 *hrs.* and produce each of their gravitational acceleration. Super gravity such as that of a White Dwarf Star or that of a Black Hole is caused by large stars that expand into Red Giant Stars as they near the end of their fuel. They grow outward with their fields to be 100 times larger than their normal sizes. Their fields speed up because they still try to circle them in 9.864*hrs.* The fields speed can also increase up to 100 times what it was and causes the Red Giant Star to collapse because the field's normal acceleration is increasing so many times. The field collapses with the star as the star becomes a White Dwarf Star having a field which maintains the speed it had circling the 100 million mile Diameter Star and now circles the tiny white hot star with a 20,000 mile diameter. The field causes the new star to make 1000's of revolutions per second because of it. Because the field's velocity is not driven by the natural 9.864*hrs* per revolution it slows over the years. But for hundreds of years the tiny star has a giant gravity field which has many times the strength it had when it was a full size star. The super velocity of these gravity fields produces the huge normal acceleration which makes the super gravity of White Dwarf Stars and of Black Holes.

To compute the Sun's fields normal acceleration ($a_n = \frac{v^2}{r}$) at its surface requires finding the distance around the Sun at its equator and then dividing this distance by the time that the field takes to make the orbit (9.864 *hrs.*). This determines what the Sun's fields speed or velocity is there.

The distance around the Sun is its circumference and it is equal to
$C = \pi D$.

$$\begin{array}{l} \quad\;\;\pi \qquad\quad \textit{Dia.} \\ C = 3.14 \times 868{,}000 \textit{ miles} \\ C = 2{,}725{,}525 \textit{ miles} \end{array}$$

Distance around the sun in miles.

$$Velocity = \frac{distance}{time} = \frac{2{,}725{,}525 miles}{9.864 hrs} = 276{,}000 mph$$

Speed of the Sun's field at its surface.

Normal acceleration of the Sun's field (Inward Toward) the Sun's center

$$a_n = \frac{v^2}{r} = \frac{276{,}000 mph \times 276{,}000 mph}{434{,}000 \textit{ (radius of Sun in miles)}} = 176{,}000 m/hr^2$$

$$a_n = 176{,}000 \text{ miles}/hr^2$$

All the average distance to the planets from the center of the Sun is known and their orbital speeds are also known. By knowing this the normal orbital accelerations of each planet can be computed by

$$a_n = \frac{v^2}{r} = \frac{\textit{velocity of each planet in orbit squared}}{\textit{distance to center of the Sun}} \quad \frac{mph}{miles}$$

Earth's orbital $a_n = \dfrac{66{,}700 mph \times 66{,}700 mph}{93{,}000{,}000 miles} = 48 m/hr^2$ (47.8)

This can also be computed from using the Sun's own a_n as it can be done for each of the planets. This proves that the Sun's field orbits it in 9.864*hrs.* because the normal acceleration of its field is computed by using 9.864*hrs.*

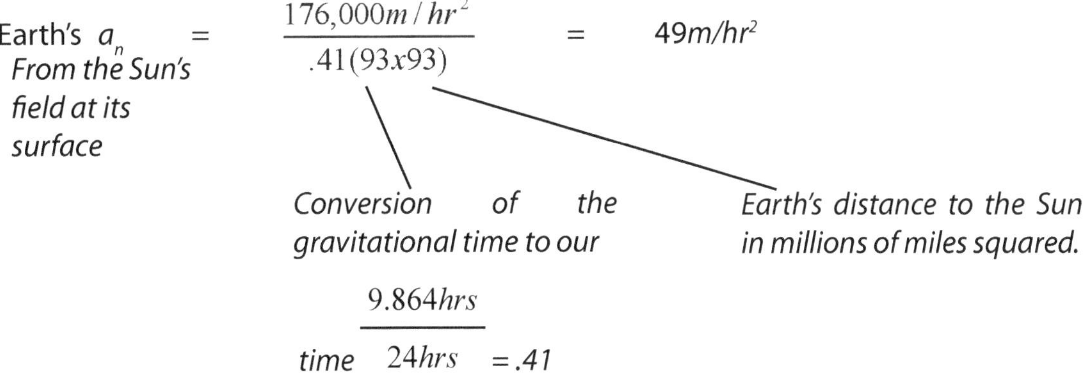

The .41 conversion is what produces Keplers constant.

Each planets orbital a_n can be computed from the Sun's $a_n = 176{,}000m/hr^2$. (See page 56) This proves that the Sun's field orbits it in 9.864*hrs.* All of the planets moons orbital normal accelerations can similarly be computed by using the 9.864*hrs.* for the time that each planets field orbits its surface at its equator.

MOON'S a_n FROM PLANETS FIELD AT SURFACE

It is also possible to compute each moon's orbital acceleration from their planets fields (a_n) using the time of orbit at their equators of 9.864*hrs.* which proves exactly what Jupiter has been telling us for millions of years.

Each star, planet and moon's a_n is produced by their fields orbital time of 9.864*hrs.* or a *Lucy*.

The preceding is the awesome proof of what causes the force of gravity and how it actually works.

Gravity no longer has to be so mysterious as to be composed of dark energy and dark matter. The force of gravity operates exactly like all other forces do, by causing equal but opposite reactions. All forces obey Newton's $F = ma$ formula. It's not only true here but everywhere in the Universe.

The action of the outer stars traveling as fast as the inner stars in some galaxies has caused some to question these truths.

Our Sun is a single star and powers our solar system. In the galaxies all the stars fields inside the orbits of outer stars power the outer stars.

All of their forces need to be used to arrive at the cause of the motions of the outer stars and not only the force of those stars at the center of their galaxies.

The galaxy's fields of gravity are aided by every star in them. The fields all meld together to produce the galaxy's field of gravity.

NOTES

1 World Book Encyclopedia, 1979
2 Morris Kline, *Mathematics and the Physical world,* P. 170
3 Ferris, Timothy. The Whole Shebang. 1997
4 Ferris, Timothy. The Whole Shebang. 1997.
5 Ferris, Timothy. The Whole Shebang. 1997
6 New Solar System, The. Edited by J. Kelly Beatty, Brian O'Leary, Andrew Chaikin. Published by Sky Publishing Corp. 1981.
7 New Solar System, The. Edited by J. Kelly Beatty, Brian O'Leary, Andrew Chaikin. Published by Sky Publishing Corp. 1981.
8 New Solar System, The. Edited by J. Kelly Beatty, Brian O'Leary, Andrew Chaikin. Published by Sky Publishing Corp. 1981.
9 New Solar System, The. Edited by J. Kelly Beatty, Brian O'Leary, Andrew Chaikin. Published by Sky Publishing Corp. 1981.
10 Jastrow, Robert. Red Giants and White Dwarfs. 1979
11 Jastrow, Robert. Red Giants and White Dwarfs. 1979
12 Jastrow, Robert. Red Giants and White Dwarfs. 1979.
13 Jastrow, Robert. Red Giants and White Dwarfs. 1979.
14 Ferris, Timothy. The Whole Shebang. 1997.
15 Asimov, Isaac. Aisimov's Guide to Science. 1972.
16 Asimov, Isaac. Aisimov's Guide to Science. 1972.
17 Asimov, Isaac. Aisimov's Guide to Science. 1972.
18 Asimov, Isaac. Aisimov's Guide to Science. 1972.
19 Ferris, Timothy. The Whole Shebang. 1997.
20 Asimov, Isaac. Aisimov's Guide to Science. 1972
21 Ferris, Timothy. The Whole Shebang. 1997.
22 Asimov, Isaac. Aisimov's Guide to Science. 1972
23 Asimov, Isaac. Aisimov's Guide to Science. 1972
24 Ferris, Timothy. The Whole Shebang. 1997.
25 Ferris, Timothy. The Whole Shebang. 1997.
26 Ferris, Timothy. The Whole Shebang. 1997.
27 Ferris, Timothy. The Whole Shebang. 1997.

About the Author

Dean E. Walker is a professional Engineer. He graduated from Norwich University in Northfield Vermont in 1960 with a Bachelor of Science Degree in Civil Engineering.

While studying at the University he was shown how to compute the effects of gravity's force. But, he was also told that no one knew what causes gravity's force. That thought never left him and he grappled with the solution of the cause of gravity ever since then. He could not leave it alone until he had read and studied everything about it and thought it out and solved the mysteries of gravity's cause. He offers his book True Gravity and the Blueprint of the Universe as proof of gravity's cause.

www.ingramcontent.com/pod-product-compliance
Ingram Content Group UK Ltd.
Pitfield, Milton Keynes, MK11 3LW, UK
UKHW060120300726
14090UKWH00002B/281
* 9 7 8 1 4 1 8 4 4 1 9 9 9 *